Remote Sensing for Environmental Data in Albania:
A Strategy for Integrated Management

NATO Science Series

A Series presenting the results of activities sponsored by the NATO Science Committee. The Series is published by IOS Press and Kluwer Academic Publishers, in conjunction with the NATO Scientific Affairs Division.

A. **Life Sciences**	IOS Press
B. **Physics**	Kluwer Academic Publishers
C. **Mathematical and Physical Sciences**	Kluwer Academic Publishers
D. **Behavioural and Social Sciences**	Kluwer Academic Publishers
E. **Applied Sciences**	Kluwer Academic Publishers
F. **Computer and Systems Sciences**	IOS Press

1. **Disarmament Technologies**	Kluwer Academic Publishers
2. **Environmental Security**	Kluwer Academic Publishers
3. **High Technology**	Kluwer Academic Publishers
4. **Science and Technology Policy**	IOS Press
5. **Computer Networking**	IOS Press

NATO-PCO-DATA BASE

The NATO Science Series continues the series of books published formerly in the NATO ASI Series. An electronic index to the NATO ASI Series provides full bibliographical references (with keywords and/or abstracts) to more than 50000 contributions from international scientists published in all sections of the NATO ASI Series.
Access to the NATO-PCO-DATA BASE is possible via CD-ROM "NATO-PCO-DATA BASE" with user-friendly retrieval software in English, French and German (WTV GmbH and DATAWARE Technologies Inc. 1989).

The CD-ROM of the NATO ASI Series can be ordered from: PCO, Overijse, Belgium

Series 2. Environment Security – Vol. 72

Remote Sensing for Environmental Data in Albania: A Strategy for Integrated Management

edited by

Manfred F. Buchroithner

Dresden University of Technology,
Institute for Cartography,
Dresden, Germany

Kluwer Academic Publishers

Dordrecht / Boston / London

Published in cooperation with NATO Scientific Affairs Division

Proceedings of the NATO Advanced Research Workshop on
Remote Sensing for Environmental Data in Albania: A Strategy for Integrated Management
6 to 10 October 1999

A C.I.P. Catalogue record for this book is available from the Library of Congress.

ISBN 0-7923-6527-5 (HB)
ISBN 0-7923-6528-3 (PB)

Published by Kluwer Academic Publishers,
P.O. Box 17, 3300 AA Dordrecht, The Netherlands.

Sold and distributed in North, Central and South America
by Kluwer Academic Publishers,
101 Philip Drive, Norwell, MA 02061, U.S.A.

In all other countries, sold and distributed
by Kluwer Academic Publishers,
P.O. Box 322, 3300 AH Dordrecht, The Netherlands.

Printed on acid-free paper

TABLE OF CONTENTS

PREFACE

From October 6th to 10th 1999 a NATO Advanced Research Workshop (ARW) on "Remote Sensing for Environmental Data in Albania - A Strategy for Integrated Management" took place in the Palace of Culture in Tirana, Albania's capital. It would take too long to recall here in detail how this event finally came about. (Some aspects are, anyway, given in the opening address of Ergjin Samimi. (Cf. page 9 of this volume.) Let me just briefly state that after the first submission of the initial idea to NATO it took more than three years until I, as the designated Workshop Director was, in conjunction with Brussels, able to decide that the ARW should take place in October 1999. Some two weeks after this decision had been made, however, the allied NATO airforces started their attacks on Serbia and the whole region was in a riot. A period of anxious waiting for stable - or at least secure - conditions in the Tirana area began. In close contact with the NATO Headquarters we proceeded in our preparatory work and could eventually, in due time before the summer break, confirm the announced ARW to all invited speakers and our friends from Albania and surrounding countries.

The Workshop, with its high-level scientific and technical presentations, stimulating discussions, social events before and after work and the speakers' field excursion to Kruja, Albania's former capital, with its magnificent mountain hinterland was – in everybody's opinion - a complete success.

Being the first of its kind in the Balkans, the ARW tried to show the enormous potential of both remote sensing and GIS technology for Albania, Kosovo and the whole region. Through the selection of speakers and topics we did not only succeed in covering the whole spectrum of relevant remote sensing applications but also to document it with concrete examples and link it with geo-information data bases. Lecturers from the U.S.A., Canada and Western Europe, but also from former socialist countries, guaranteed the workshop participants a good mix of experiences. The availability of demonstration facilities for digital data sets and software, together with the excellently prepared venue at Tirana's Palace of Culture also contributed to the high degree of efficiency of the lecture series.

The Workshop was directed towards a wide spectrum of participants reaching from advanced students in environmental, social, engineering and geo-sciences via employees of technical bureaus and administrative authorities up to decision makers on industrial management and ministerial level. The "Manifesto of Tirana" which was jointly phrased and unanimously adopted by lecturers and participants on the last day of the ARW outlines a possible strategy and master action plan for the implementation of remote sensing and GIS technology in Albania. This request for urgent implementation was subsequently sent to the President and the Prime Minister of Albania and to all relevant ministries. I hope that it will be translated into action soon. This would then mean a long-term "sustainability" of the NATO ARW "Remote Sensing for Environmental Data in Albania - A Strategy for Integrated Management".

Dresden, Germany, May 2000

Manfred F. Buchroithner
Workshop Director

ACKNOWLEDGEMENTS

As Workshop Director and Editor of this volume I should like to convey my gratitude to NATO and their initiative "Science for Peace". The Director of the NATO Environmental Programme, Dr. L. Veiga da Cunha, and his successor, Dr. Alain H. Jubier, kindly and strongly supported the whole undertaking until the delivery of this book. Meeting with Dr. Jubier in Tirana was a unique experience and pleasure for me.

Without the initiative and Workshop Co-Directorship of Prof. Dr. Ergjin Samimi the whole event would never have taken place. My humblest thanks go to him for all his assistance, encouragement and friendship.

All the logistic support on the spot came from Mr. Ilir Dedei, hotel owner, landlord and businessman in Tirana. He managed everything from the airport shuttle service, our accommodation and food to the joint field trip, and even fulfilled rather extravagant wishes. In the preparation phase and during the symposium he became more than just a "project partner".

Our "liaison officer for east relations" at the Dresden University of Technology, Dr. Günther Krause, was my "right hand" during the complex organisation of all the lectures. He was assisted by Matthias Höfner, one of my students, whose major merit, however, was the realisation of all the editorial "legwork" which was necessary to make this book appear in a somewhat consistent and homogenous structure. Whoever edited a volume of 254 pages with 17 different contributions knows how much gratitude I owe Matthias. Thank you!

Last, but not least the smooth co-operation with Kluwer Academic Publishers has to be gratefully acknowledged.

Manfred F. Buchroithner
Workshop Director

OPENING ADDRESS BY DR. ALFRED MOISIU, PRESIDENT OF THE ALBANIAN ATLANTIC ASSOCIATION

Ladies and Gentlemen,

It is a pleasure and honour for me as President of the Albanian Atlantic Association to address the speakers and participants of this workshop concerning some contemporary scientific problems related to a more vigorous and prospective development of Albania and the entire region.

It is true that Albania is a profoundly European country and inhabited by a people ranking among the most ancient on the Balkan Peninsula, their origin deriving from the Illyrians. Unfortunately, however, the historic events and developments have made our country one of the poorest and least developed in Europe. But the desire of our people and especially of the youth of Albania is that we must continue to be part of Europe not only geographically. We must be an active member in all fields, including the most advanced scientific ones. It is not my duty here to teach you how important this is. I would, however, like to tell you that just like people are supposed to know a common language to improve their communication, they should also know and exchange the newest information in science, thus paving the road towards progress.

Albania, inhabited by a people who uses unique language, encircled by neighbours using quite different languages, also with a different history, different traditions and culture, has had a tragic destiny. Although known as a militant people, the Albanians have never been aggressive throughout their history. They have never attacked others, but unfortunately always been victims of foreign aggression and, naturally, in these cases they have tried to respond to these aggressions. I cannot say whether this has been negative or positive for the Albanian people, but it is a fact. Only within this century, Albania has been turned into a battlefield several times, in two Balkan Wars in 1912 and 1913, in World War I and World War II.

Albania has always been a crossroad. The Byzantine, Roman, Ottoman, Austrian-Hungarian empires have passed through its territory. All these have left their positive and negative traces on Albania's development.

Unfortunately, the ruin of the Ottoman Empire found Albania without powerful friends. Hence, beginning with the Berlin Congress in 1878, the London Conference in 1913 and the agreements after World War I, Albania was truncated and deprived of its most fertile, richest grounds and underground parts. In this way, of whole Albania which, according to some foreign scholars, covers an area of about 80 000 km², remained only 28 000 km². We may say that Albania is the only country in Europe, which along all its state frontiers is bordered with its compatriots. There are nearly 7 million Albanians in the Balkans, of whom only half live in Albania.

It is understandable that after all these unjust actions, the Albanians experienced a series of opposite events. To his end, Albania became a rather important factor for the

1

M.F. Buchroithner (ed.),
Remote Sensing for Environmental Data in Albania: A Stragegy for Integrated Management, 1–6.
© 2000 *Kluwer Academic Publishers. Printed in the Netherlands.*

peace and stability in the Balkans. It is common knowledge now that the destruction of Yugoslavia started in Kosovo in 1981 and later in 1989, when Milošević curbed Kosovo of its autonomy of 1974. But since the very beginning, it has also been clear that this destruction would end in Kosovo as well. Although by now the war in Kosovo is over, still time and work is needed to completely calm-down the region.

What has been attained is a result of the efforts of the Albanian people of Kosovo. But it must be openly admitted that the Albanians would not have been able to solve the problem based on their own forces alone. If Albanians were left alone to face the Serbs, the war would not only have been longer, but also mankind would have had to face even more serious crimes, and the situation would have aggravated to an extent that it might have led to the burst of a war in the Balkans and broader. It is a known fact that a great number of Albanians are living in Macedonia, in Montenegro and in other countries. Consequently, the outbreak of a very long war between Serbs and Albanians would have involved all these countries.

It is clear that one day the injustices would be laid for solution. The Serbs elaborated a first theory on the elimination and assimilation of the Albanians in 1844, described in "Naçertanija" of Ilia Garashanin. But the Albanians, first of all, opposed this inhuman and colonial theory with the demographic "weapon". Hence, although again and again the Serbs violently expelled hundreds of thousands of Albanians and massacred and eliminated thousands of others, still the Albanian population kept growing.

It is an undeniable reality that the crisis was solved by NATO, the U.S. and the European Union. Without their decisive, military and diplomatic aid and interference, the crisis would have continued too long, claiming many sufferings and blood of innocent people. The Albanians will never forget this. NATO demonstrated their value as a powerful political-military organisation, which after the end of the cold war, is, along with other things, taking a new dimension with a new strategy, as a defender of democracy, human values and the values of European civilisation. Likewise, the U.S. and Europe clearly showed that they stand in defence of the U.N. principles not only in theory but in practice as well.

International diplomacy proved it has started to draw lessons from the events in Bosnia. The political, diplomatic and military operations we saw, showed that the time when every regime may do what it wants with its own nationals - although it has accepted and signed the international documents on the human rights, on national minorities and for the protection of democratic values - has come to an end.

Europe cannot make a progress without strictly respecting and implementing the norms it has set to ensure freedom, equality and normal development. Every country must completely observe the accepted norms, otherwise it will be reminded of these norms and finally be forced to implement them. These norms will be on the basis of the security and stability for the coming century.

We cannot say that the Kosovo crisis has been solved completely. Still there are hostilities and mutual persecutions existing between Albanians and Serbs. Certainly the Albanians have it difficult to forget the sufferings and humiliations made to them for nearly one century, especially the events of 1998 and 1999, until the land forces intervened in Kosovo. You all know this because it has been shown on TV.

Likewise, it is not easy for the Serb minorities to change from the position of a patrician to an equal position.

Kosovo has been greatly destroyed, especially in the countryside. It is common knowledge now that about 70% of the houses in Kosovo have been torched and ruined. Their construction needs time and sweat, not to mention the need to replace the ruined furniture, the destruction of agriculture and livestock.

The European assistance pledged for this purpose is being delivered rather slowly. Winter is coming, many people are still living in tents. Nonetheless, we must admit that the Albanians of Kosovo are resolute and aware that they will have to continue their sacrifices to reconstruct Kosovo, its economy and infrastructure, to restore law and order, and to build the administration and other things. The Kosovar politicians, despite the differences they have, share the opinion that Albanians will reconstruct Kosovo. Others will only assist. They are trying to reduce as much as possible and as soon as possible the ethnic tensions. Now Kosovo is a country without law and order, without administration, without police, but still with an absolute rule and tranquillity. All the people are working. At a first glance it seems somewhat paradox but the reality is the one I told you. I was in Kosovo ten days ago. With my own eyes I have seen what I already told you. I met with most of the political leaders there and with many people of various social strata. They all shared the same opinion and manifested the same determination.

Certainly, along with economic problems, there go still many rather sharp political problems. The status of Kosovo is not solved yet. Although Kosovo is de jure considered part of Yugoslavia, all know and understand that it is very difficult - not to say impossible - to turn back again to that framework. And this is what all Albanians, without exception, think. But not only Albanians, after all the events that followed, many foreign analysts cannot see a future for Kosovo in that community.

There can be no other alternative when all other republics, with Slav populations, have seceded, when Montenegro, too, is seeking to separate. Without mentioning other facts, only based on what I said above, it would be utopia and somehow unimaginable that a people with quite a different language, different customs, culture and history can be part of the remaining state system of Yugoslavia.

I believe that the international organisations operating in Kosovo will play a special role in accelerating the solution of the problems. First of all, justice must be settled, that is, the criminals who have made crimes in Kosovo must be sent to court. Secondly, it is important that as soon as possible conditions for new elections must be created. Thirdly, it is very important for the country's economy to start functioning. I have the conviction that the Albanians of Kosovo are willing and working so that Kosovo can as soon as possible become a factor for stability in the region. However, the question arises, what will be made with Serbia? The aim is to *turn it into a democratic country* as soon as possible. But will this happen soon? I think no. Why?

First - Only the elimination of Milošević from power cannot lead to the democratisation of Serbia. This is indispensable but rather insufficient. Unfortunately, since their childhood, the Serbs are educated with the spirit of racism, nationalism and chauvinism. Hence, a sort of brain-washing is needed so that they have a right judgement and accepted as unavoidable the changes that happened in Kosovo and in former Yugoslavia in general.

Second - Miloševic cannot be removed from power easily. He still has a powerful support in the people (especially among extreme nationalists and in the countryside), he has full control on propaganda, police and army. The latest demonstrations, which have not been so powerful, and the reaction of the police towards them are the best proof of this.

Third - The Serb Army still represents a very dangerous potential for the stability of the Balkans. I may say without hesitation that, although we do not have complete conclusions reached, the army has remained almost intact by the military campaign in Kosovo, without considering here the aspect of its moral. But this too, I think, can be compensated with the experience they gained by confronting the largest military power the world presently has, NATO, although the confrontation was in the air and not on the ground.

Fourth - Russia is of no less importance here. It does not want Miloševic to give up the power. On the contrary, in spite of some very formal statements, it is seeking that Miloševic sticks to power as much as possible. They would consider the toppling of Miloševic as an unacceptable loss. It would open the way to the NATO influence to extend throughout the Balkans, something which the Russians will never wish and cannot accept.

There are many *opinions* circulating *with regard to* the issue of *Kosovo, contesting its independence*, and I am mentioning only two of them:

First - If Kosovo gains its independence, after some time this may lead to its unification with Albania, thus creating a "Great Albania". Such an expression sounds to me more comical than political. After the fall of the Berlin Wall, the two German states united and as we have seen: nothing has happened. A 5 - 6 million-Albania cannot pose a threat to the Balkans and Europe. This comes to show that thanks to their political maturity, the European peoples do not at all think of a separated economic and history development, but they see it now at a united level.

Hence, Albania has its own size, but it can never be considered big and, consequently, cannot be a danger to anyone. This thesis is merely a hypothesis concocted by those who wish Albania ill and who want to exploit the blessings of the Albanian land as long as possible.

Second - The independence of Kosovo will destabilise Macedonia. Certainly, there is no truth in that. The fact that during the war in Kosovo nearly 300 000 Albanians went to Macedonia and nothing happened there, shows that Albanians do not want to cause the destabilisation of that country, they have always been and remain for a peaceful solution of the pending problems through democratic ways. Certainly, the leaders of the Macedonian policy must be more mature and human. The events of Bllaca proved the opposite. In a state where it is officially accepted that the Albanian population consists of 23% of the whole population of the country, and in fact it is as high as 45%, it cannot be allowed that these people be considered as second-hand citizens for a long period of time. These problems demand solutions, but always within the revised legislation of the Republic of Macedonia.

I just mentioned two of the most spread versions, but I cannot say that these are the only ones. However, now we are all understanding that we cannot continue in the 21st century with the mistakes made at the beginning of the 20th century.

At the end of my speech there remain still some other *problems, related with Albania* untouched, not because they are not important, but I think that you know already a good part of them or you will come to know them during your stay in Tirana through the contacts with intellectual Albanians. Nevertheless, I am mentioning here some of the most important issues.

First, I would like to tell you that the problems of the Albanians could not be solved without, first of all, the conclusion of the Kosovo Crisis.

These problems have continuously influenced in the inner-Albanian developments. The events of 1997 were the prelude of the events in Kosovo. The ideators of the anti-Albanian campaign in Kosovo and their supporters were interested in that Albania be as weak as possible, with the least influence during the pre-planned crisis. They attained this. It is well-known what in 1997 happened in our country. The consequences can still be seen.

Second, Albania took the road of democracy after the fall of a very savage dictatorship, amongst the most savage in the former socialist cam. These events found the country without competent politicians, without an elaborated strategy on how to pass from one system to the other, with extreme poverty and completely isolated. Under these conditions, Albania started an accelerated transition period, with not much order. Economy started to register inflation, a very necessary privatisation began but it was not well-structured, real estates were not restored to the former owners, the pyramid investment schemes flourished a lot and came out of control, emigration assumed unprecedented proportions, etc. But, until the end of 1996, Albania seemed to have found its way. However, the events of 1997 made many weak points and mistakes of the transition period evident and carried the country backwards.

Thanks to the care manifested by Albanian politicians, in March 1997 a civil war could be avoided. But under these conditions, organised and ordinary crime, corruption, contraband and destruction assumed unseen magnitudes. Such a situation created quite a different view and impressions which were not imagined before. Under the present government, recently these phenomena are changing, although not at a very fast pace.

Under these conditions, during the Kosovo Crisis, Albania accepted, accommodated and was a sanctuary for more than 500 000 displaced Kosovars. What was more interesting and unexpected was that 70% of the displaced were hosted by Albanian families. Before international organisations got the situation under control, the people themselves managed the situation. This was really a surprise for the European countries, but not for the Albanian hospitality.

What are the most pressing problems for Albania today:
- public order
- unemployment (which amounts to over 30%)
- lack of investment
- weak infrastructure
- corruption
- the need to restore hope in the people, particular in the youth.

Work is going on but Albania cannot change the present situation without international support. The Stability Pact is a rather interesting and promising solution. It is known that the Albanian factor in this pact is very important. We hope that this German proposal adopted by Europe and the U.S. will yield its fruits as soon as possible.

The Albanian Atlantic Association, which has so far held four international conferences on the problems of security in the region, plans to hold its 5th conference in November this year on the topic "For a Peaceful Balkans in the Coming Century". Certainly, in the focus of this conference will be the main ideas of the Stability Pact.

I believe the foreign scientists participating in this NATO Workshop need to know some problems of this character in order to intensify their contribution to the solution of specific problems, in the interest of Albania, the peace and security in the region, which is a basis for the future and for a long-lasting peace throughout Europe.

Scientific aspects represent one of the directions treated by NATO. They aim at ensuring a better and greater understanding among NATO members and their partner countries and to consolidate the stability in the Balkan region and in Europe. The organisation of this Workshop in a country where NATO is present, in a region where NATO has been operating for nearly four decades and nobody knows how long their necessary and useful presence may still be needed, shows that this event is not only of interest for scientists but also for the leaders of the Alliance.

I wish the Workshop all the success, and I also wish you to enjoy your stay in Albania.

OPENING ADDRESS BY DR. MYSLYM PASHA, DIRECTOR OF THE ALBANIAN MILITARY TOPOGRAPHIC INSTITUTE

Honourable Professor Buchroithner, Ladies and Gentlemen,

As the Director of the Albanian Military Topographic Institute, an organisation which is responsible not only for military purposes but for all official topographic mapping activities in Albania, I would like to express my deep thankfulness to the Co-Chairmen, for the invitation to participate in this workshop, co-organised by representatives of the Institute for Cartography of the Dresden University of Technology and of Alb-Euro Consulting, Tirana.

At a first glance it seems that the addressed topic is not appropriate for the recent situation of Albania where most pressing problem is how to face the challenges in passing this harsh transitory period, with many difficulties entering the open society and the free market economy.

A "classical", old one mentality prefers terrestrial work and only then thinks of remote sensing, which - as many stress - is not of need of Albania. We all here know that this is completely unreasonable and backward oriented. So this workshop improves not only scientific knowledge, but proposes contemporary and at the same time progressive methods which, combined with other geoinformation recourses, should have significant impact on the acquisition of environmental data in Albania for fundamental development projects.

This workshop, for me and my organisation, takes place at the right time and in an appropriate situation to inspire our progress and prosperity.
Let me briefly touch upon five issues:

First - MTI, the Albanian Military Topographic Institute, is very much interested to maintain and influence this activity which opens a window to modernise the present methods for map updating with new, detailed information for this land which hopes for a developed society and a modern infrastructure.

MIT as the Military Survey has still another duty, very important indeed, i.e. to review many of our procedures and requirements for the support of up-to-date geoinformation for military activities and for European and worldwide operations to maintain multinational peace support.

In the military field the Kosovo Crisis was a brilliant example how to us air- and spaceborne imagery.

Second - Many in this audience are very familiar with the activities in the field of geoinformation in Albania. To make this topic clear, I would like to stress that we should act with converging goals in order to work for the implementation of a national strategy for the integrated management of geoinformation resources.

M.F. Buchroithner (ed.),
Remote Sensing for Environmental Data in Albania: A Stragegy for Integrated Management, 7–8.
© 2000 *Kluwer Academic Publishers. Printed in the Netherlands.*

We are fortunate to welcome her in Tirana representatives of internationally renowned institutes of remote sensing and to share their experience, which is very useful for us. So, geographers, geologists, environmental experts, road designers, land-use policy and land registration managers in governmental and private agencies hope to have a reciprocal and goodwill co-operation toward the new ways of remote sensing possibilities.

Third - As mentioned before, the democratic era, the "wind of changes" shook up the whole society and with them all cartographic institutions, on their way to the European integration.

Unfortunately, the whole spectrum of mapping activities is developing not according to a strategy but on an emergency basis. The challenges of this transition affect various agencies which obtain, collect and maintain geoinformation for the public and private sector. But there is a risk: if different segments favour different technologies, they should have big problems and useless investments may be made.

The Project Management Unit for Land Registration, e.g., has actually the potential to introduce new technologies in photogrammetry and other disciplines. So our particular needs should be by this unit. If a remote sensing project materialises - and we all hope this obviously - the advantages and possibilities of this technology should be open for all agencies.

Fourth - With support by foreign experts present during this workshop, we should convince our government and politicians of our needs and ask them to support us. A Remote Sensing Centre in Albania is one of our dreams.

We do not only like to see our globe, with brilliant colours, in spaceborne imagery but to use them for mineral, river, road, pollution, bridge, temperature, etc. mapping and monitoring, and to develop strategic plans for our progress.

Finally, MIT with my humble contributions, is ready to assist in all initiatives and to support any co-operation with its resources in geodesy, mapping, logistics, etc.

I would once again like to express our deep thanks to the remote sensing experts from Europe and America for their contributions in this workshop.

OPENING ADDRESS BY PROF. DR. ERGJIN SAMIMI, ALB-EURO CONSULTING TIRANA, WORKSHOP CO-DIRECTOR

Ladies and Gentlemen,

Let me first of all tell you a few words about history of the workshop which we are opening today.

Four years ago Dr. L. Veiga da Cunha, Director of the NATO Environmental Programme visited Albania. At that time I was preparing an application for a NATO workshop which I gave him during a meeting. At first he was very sceptic at first but after some phone calls and meetings he accepted my application and appointed Prof. Dr. Arthurthon from the British Geological Survey as Director of the envisaged workshop.

Then the Kosovo Crisis started and everything became very unsafe here in Albania. Prof. Dr. Arthurthon got retired.

My second proposal for the position of the Workshop Director, which was accepted by Dr. L. Veiga da Cunha was Prof. Dr. Buchroithner from the Dresden University of Technology. His staff and efforts we have to owe that specialists and leaders in the fields of remote sensing and GIS are now here in Albania among us. Thank you very much for accepting the invitation to come to Albania.

I would like to mention that for the first time a workshop with an Albanian and foreign audience takes place in Albania. About 15 experts will show us how to enter the new era of remote sensing and GIS here in Albania.

On the other hand, we have the participants from all fields of science. They are decision makers among whom you can find people responsible in the fields of agriculture, hydrometeorology , water/land problems, forestry/environmental problems, infrastructure and geography, people of science and culture etc. They are the "students" who are willing to learn everything new presented in the next days.

I am sure they do not pretend to call themselves specialists in remote sensing and GIS at the end of this workshop but they will have got an overview of what is possible using these technologies. They will know what they can use in their special fields to increase the quality of products. A first step will be done with this workshop.

In the third part of my speech I would like to talk about the current situation of remote sensing in Albania. It will show you that it is possible to introduce a new technology in a very short time in a cost-effective and qualitative way.

Concerning satellite images we in Albania have a basic knowledge of what can be done with them. I want to mention the CORINE Land Cover Project which was started at the Geographic Studies Centre and finished at the Soil Institute with the conventional interpretation of the images at the scale 1 : 100 000 using standardised European methods. This was followed by a pilot study on land use in a commune and some samples of soil.

9

M.F. Buchroithner (ed.),
Remote Sensing for Environmental Data in Albania: A Stragegy for Integrated Management, 9–12.
© *2000 Kluwer Academic Publishers. Printed in the Netherlands.*

So far, this is all we have done in the fields of remote sensing and GIS. Some Albanian entities perform field work to cover map and plan needs in different regions and jobs. But this work is not based on accurate geodetic basis and up-to-date data.

The MIT is much esteemed, last not least for its preparation of the Topographic Maps, at scales from 1 : 10 000 and 1 : 1 000 000, which were at the beginning prepared on the basis of Russian aerial photogrammetry. Later they used Chinese and now they are using American imagery. The maps mentioned cover the whole of Albania and were a good basis for every kind of studies and projects we were doing.

Some other works in these fields have been done by the Institute of Geo-Topography which makes plans for Albanian cities at scales of 1: 500, 1: 1 000, 1: 5 000. Moreover, some private companies tried to cover the construction work. The Institute of Topography was forced to stop its work for a short time because of a lack of money.

The implementation project unit PIU of the land registration uses many different data sources for property registration, infrastructure etc. It is financed by USAID and the EC. For their work they use 7000 km² of aerophotogrammetry, processed at analogue, analytical and some new total stations as well as GPS.

Because of their reorganisation the Ministry of Public Work is presently doing nothing in these fields. Whenever Albania gets a new government, one of the first steps is the complete change of the technical management.

As you can see our situation in the field of land information is not the best. It is too weak, not accurate enough and, most important, not up-to-date. At the moment we are using technology of the 1960s. We know that land information is the basis for all other units dealing with information, and keeping in mind that the next century will be the "century of information " there is one thing we should start really soon: to develop a strategy for the future.

The basis would be the foundation of a Laboratory of Remote Sensing and GIS. Albania needs to move forward fast, but its information system is staying on the same level. The development of the western world is supported by the technology of information. One of the most expressive information technologies which

- link the environmental information, economy, culture, society etc.
- perform analyses of the influence of spatial phenomena
- do the analyses of influence of spatial phenomena
- determine the priority of regions
- say *what* must be planned and done and *where*

is the technology forming the basis of a new infrastructure. It is called Remote Sensing / GIS Infrastructure.

This technology is based on satellite and conventional data offering us the most effective ways to create the bases of the other infrastructures to achieve the most required one:

"Sustainable Development within a Protected Environment"

In this case we have to

- implement multi-purpose remote sensing /GIS systems as well as specific ones; middle- and long-term job qualifications
- implement training possibilities for young people
- thus give Albania and its decision makers possibilities and solutions for economic, environmental and social problems.

If this does not happen in the near future, Albania will be one of the least developed in that field.

I would like to stress the following aspects.
"The Future is now!" Albania should move fast to build up its future. All speakers are welcome to give their presentations and to discuss for the benefit of Albania. Please lead us to find the way.

12

Appendix

Implementation of a Remote Sensing and GIS Laboratory (an example)

Tasks: - creation of a digital database
          - education and qualification
          - thematic design
          - analysis of problems depending on many different factors
          - application of new technologies
          - design and implementation on central, provincial and prefecture level

Costs (as of summer 1999):

Equipment	1000 US $
Hardware	150
Software	350
Peripherals	160
Job training	82
Consumable	24
Images	454
Auxiliary etc.	282
	1500

Space: 150 m²

Employment:

Specialist	11
Auxiliary .	4
	1500

Time of Implementation: 6 – 9 months

ENVIRONMENTAL PROBLEMS OF ALBANIA

PERIKLI QIRIAZI
Department of Geography, Tirana University
SKËNDER SALA
Geographic Studies Centre
Tirana - Albania

1. Account of the Physical, Political and Economical Setting of Albania

1.1. NATURAL SETTING

Albania is a Mediterranean country with an area of 28 748 km². Within its borders you can find different landscapes: the most western part, bordering the Mediterranean Sea, comprises hilly areas and recent and historic glacial plains in different elevations with extensive agricultural use of the Mediterranean type, while more towards the east the continental character of climate is more pronounced. The major part of the hinterland is mountainous or even alpine in character, showing among rugged parts also high-elevation plains and in small spots even perennial snow. The surface is a mix of dense forest and exposed rock outcrops on steeper slopes. Because of this diversity of geographical landscapes and their biodiversity Albania is called the "Great Natural Museum".

The Albanian territory is *geologically* composed of sedimentary formations, especially terrigenous ones (flysch and molasse) and limestone, as well as magmatics - especially ultrabasic and metamorphic formations. Its geological structure is complicated by old and new tectonic detachments. This geological structure and a long morphotectonic and morphoclimatic evolution has affected the ground treasures: there are 40 kinds of raw materials, among them ores of Fe, Ni, Cr, Cu, natural gas etc.

In Albania's *relief* predominate hills and mountains (over 76%), with a high level of relief energy. Agricultural fields occupy small areas and lie mainly in the western part. Agriculture has caused serious erosion.

Albania has *Mediterranean climate* with an obvious modification in the vertical direction. The average temperatures of January change from − 3° to 10°C, in July from 17°C to 26°C. In average daily means of 10° are exceeded during 156 days per year. The average precipitation amounts to 1480 mm per year. It has an irregular regime and geographic distribution. This causes aridity during the summers and frequently even in the other seasons, and high humidity during the winters.

Hydrography is, on the surface, manifested by a very dense drainage network with a total length of 49 027 km, which in average discharges 1308 m³/sec resulting in a total annual volume of 41.2 km³ water, from which every Albanian inhabitant gets 14 000 m³

M.F. Buchroithner (ed.),
Remote Sensing for Environmental Data in Albania: A Stragegy for Integrated Management, 13–30.

and each km^2 of surface 1,433,000 m^3. The solid freight amounts to an average of 1,650 kg/sec, resulting in a total of 60 million tons.

Albania has big tectonic lakes (Shkodra, Ohrid, Prespa), but also carstic and glacial lakes and coastal lagoons. The low degree of accumulation of the water reserves, mainly during the cooler period of the year, creates many problems for the economy. In some regional centres there exists a lack of the necessary amount of water during that period.

Flora and Fauna. Because of its wide ecological range and its geographical position within the Mediterranean Basin, Albania boasts a rich biodiversity. There is a very rich flora that contains 1. *vascular plants* (about 3200 species), which also represent 30% of the Balkans' flora; 2. *non vascular* plants (about 1800 species) comprehending 600 species of mushrooms, 500 species of *bryophyte,* and about 700 species of *liquors.* In addition, there exist a lot of plant habitats like subalpine and alpine meadows, forests (broadleaf, coniferous, and mixed), shrubs, pastures, marshes, river plains, coastal lagoons, etc. About 30 species of Albania, Greece and former Yugoslavia are endemic, and 180 species are subendemic. Some of them belong to phytosociological endemic or subendemic units, like: *Euphorbietum genistetum hassertianae, Forsythetum europea, and Peteriutum ramentacea.*

Four plant belts can be distinguished: shrubs and Mediterranean forest (maccia), oak forests, beech forest, and coniferous forests and alpine pastures. Forests cover 36% of the whole territory.

Although there exists no complete knowledge about the faunistic provinces of Albania, we can at least state that it shows a great diversity. We find 70 species of mammipherous, 320 species of birds, 37 species of reptiles, 15 species of amphibians, 313 species of fish (4 of them endemic ones), 3580 species of insects (900 of them butterflies). Albania is a very important place for migrating birds, especially for winter fowl. There cross migration roads of some endangered species like *Numenius tennuirostris.* In Albania we also find some special species like bear (*Ursus arctos*) and wolf (*Canis lupus*) etc. which are endangered in large parts of Europe.

1.2. POPULATION

The Albanian territory has been populated since Palaeolithic times. This also means that the human impact on the environment has very long duration and, as a result, its today's consequences are rather serious. Presently Albania has 3.6 million inhabitants.

54 % of the people live in the countryside. In 1990 the urban population represented 36 % of the whole population. In the following years this portion grew as a result of a recent, motorised movement of the population from mountainous and hilly areas into the plains. This has caused a tremendous growth of the cities, accompanied by severe environmental problems.

1.3. ECONOMY

Albania boasts many natural resources. In the light of this fact we can say that the poverty if the population is not at all the result of the natural environment, its climatic factors or even drought, but it is caused by the socio-economic system.

From the period before World War I Albania has inherited a backward-oriented economy. During the first decades after this war some new branches of energy, primarily electrical, mineral, mechanical, paper, wood industry, etc. developed. Applying a short-sighted policy, the state gave priority to the heavy industry, however based on very old technology. This resulted in negative consequences for the whole environment.

In agriculture the total area has grown from 290 000 ha in 1938 to over 710 000 ha in 1989. For this purpose march land has been reclaimed. Many marches in the western lowland have been dried and new fields on hilly and mountainous slopes opened, even in slaty slopes by cutting shrubs and woods. Agriculture first mainly oriented in cereals and subsequently in industrial plants like cotton or tobacco. This re-orientation seriously damaged the original ecosystems with high ecological values and resulted in damaged plants and reduced biodiversity, intensified erosion and land degradation.

The socialist energy generation system vanished very quickly. Since the Albanian economy had been one of the most centralised and isolated ones in Europe, especially during the seventies and eighties, very soon it fell behind, without any possibility for a regeneration of its antiquated technology. All this brought the Albanian economy into a deep crisis, which reached the lowest point at the end of the eighties, where the economy had a total collapse. So it became necessary to undertake radical reforms. These reforms were only possible after the establishment of a pluralistic system, i.e. after 1991. In collaboration with foreign financial institutions Albania began to successfully apply the reforms, which are now directing the country through the free market. With foreign, Albanian and joint investments new factories are built, especially those for light and food industries. Moreover, some of the existing factories are reconstructed like chromium, copper and petroleum plants. Hydropower stations, lumber factories, factories for construction materials, etc. continue their work normally. At the same time many heavy industry plants, built during the communist period, had to conclude their activities, because they could not compete with the free market.

After the land re-ownership process following the political changes in the nineties, the farmers became more interested in land protection. Now Albania is slowly harvesting the fruits of these liberating initiatives, which had been started at the beginning of the nineties. Albania has obtained a good place in its agricultural development, which is now working under free market conditions and covering 75% of the country's needs for agricultural products. Dictated by the trade requests, the agricultural production is very quickly changing its structure. The farmers' economic welfare is determined by animal husbandry, the number of animals is increasing. In general, all these agricultural activities have a positive influence on the decrease of the erosion.

The present fully private ownership of the transport has also helped Albania a lot. The number of transport vehicles increased tremendously. Some projects for highway construction, like the so-called Eight Corridor, some for the extension of the railway transport, for the seaport and airport are under preparation. But all these however are connected with many environmental problems.

Albania offers a high touristic potential, with many possibilities for tourism development. During the communist period incomes from tourism have been very low, compared to other branches of economy. But during the late nineties, when the new development strategy for tourism was developed, the tourism concept has changed a lot. This new strategy will head towards a qualified eco-tourism. A too quick tourism development, however, might also cause environmental problems.

After 1991 a new era began for Albania. Many reforms in industry, agriculture and transport have started. The country with its natural, human and economic potential offers a good basis for a healthy tourism. After 1990 with the definition of a new strategy for tourism development the tourism concept has changed.

2. Environmental Situation

Various facts have coined the present condition of Albania's environment. These are:
- a low development of industry
- a low quantity and quality of agriculture
- no large urban remains
- Albania's surrounding by neighbourhood states with a low development of industry.

Recent measurements showed a lower level of pollution of water, air and land than the allowed norms. This is e.g. indicated by the fact that along the Albanian coast one cannot detect any algae with a bad smell.

The present environment of Albania shows traces of two political systems. These totally different systems are manifested in different environmental policies, so their impacts are not the same.

2.1. COMMUNIST PERIOD

During this period negative and positive factors have affected the Albanian environment. As the *negative factors* we can mention:
- a bad administration of natural resources, e.g. forests. In the latter case the proportion between cutting and growth has been 3 to 1.
- use of antiquated technologies in all branches of industry, especially in chemical, leather, and paper industry. They had a high level of pollution.
- a wrong orientation of industry and agriculture and a wrong geographic setting of some industrial objects which has caused environmental problems for specific areas. In these regions air, water and land are polluted beyond the permitted level, their soils are degraded, and their forests and other natural ecosystems are damaged.
- the complete absence of a legal base and state institutions for environmental protection
- no distinction between industrial and urban remain treatment
- insufficient water supply, also for remain treatment
- a low awareness of environmental problems. This is the result of the total absence of information about the environment.
- environmental impacts in other fields like bad hygienic conditions, lack of health care etc.

As *positive factors* we can mention:
  - lack of nuclear industry
  - a low level of consummation; use of returnable packing; high percentage of mass transport and use of bicycles; avoidance of certain chemicals, cosmetic products etc.

2.2. TRANSITORY PERIOD

During this period a lot of *positive factors* affected the environment:
  - As a result of the work interruption in the greatest part of all industrial objects and also of the low use of pesticides in agriculture, the environmental pollution dropped significantly. Now agriculture is becoming more natural.
  - use of electrical energy instead of other ways of warming
  - Approbation of a legal base for environmental protection, creation of governmental and non-governmental environmental organisations.
  - tendency to use new technologies in industry and agriculture, according to the international legislation for environmental protection; participation in international treaties concerning environmental problems.
  - For the first time studies are carried out regarding the evaluation of environmental problems.
  - changes in economic structure like privatisation, growth of energy price, wood, water and fuel customs policy. All these have created a better situation for the environmental protection.

Negative factors:
  - lack of proper laws and powerful institutions for environmental protection. This has created a favourable situation for damages in forests, pastures, waters, mines, etc.
  - lack of foreign investment. This creates a favourable situation for investors to bring still install old technology which pollutes the environment.
  - An uncontrolled movement of population from mountainous areas caused a degradation of urban environments and very bad hygienic conditions.
  - Replacement of returnable packings by non-returnable ones is affecting the increase of pollution.
  - intensive automobile traffic and use of fuel of low quality.

2.3. QUANTIFIACATION OF ENVIRONMENTAL STATE

In order to get an idea about the present environmental situation we will, in the following, analyse the pollution level of air, water, land, and forest as well as the level of erosion, etc.

2.3.1. *Air Pollution*
The air pollution is caused by different kinds of gases, dust and soot. During the *communist period*, the main reason for this pollution were gases, dust and soot released

by industry and urban activities. Transport, due to the low number of vehicles, had had a smaller impact.

During these years the following agentia have been released into the atmosphere: 270 000 t SO_2, 9 100 t H_2SO_4, 2 400 t CO_2, 760 t NH_3, 700 t NO_x, and 3 573 000 t dust/year. These amounts of gases, soot and dust have polluted all the areas around industrial plants, e.g. in Kukes, Rubik, Laç, Elbasan, Tirana, Fier, Vlore, etc.

During the *transitory period* the closing of most of the industrial objects reduced the amount of the released gases, dust and soot in atmosphere. Although during these years many polluted industries did not work, the percentage of the polluted elements did not decrease too remarkably. Still, we have the inherited pollution from these old industries. Within the last two years an increase of some industries with polluting effects like petrol industry (120%), coal industry (25%), etc. has to be noted. Also, the chromium industry is still working which for every ton of iron-chromium products releases about 1000 m³ gases into the atmosphere that contain SO_2, H_2S, CO_2, CO, NO_x and 300 t/year of dust.

Based on data published by the Public Health Institute (PHI) and Hydro-Meteorological Institute (HMI) one can get an idea about the present level of atmospheric pollution:

In 1998 industrial activities released about 249 200 t gases into the atmosphere, the major contributors being the chromium industry and TEC.

Transport. Urban traffic is another serious problem. Because there are not enough data available, only information for Tirana is given. In this city there exist some 33 500 vehicles

TABLE 1. Increase of vehicles in Tirana.

Year	1990	1991	1992	1993	1994	1996	1998
Number of vehicles	8 700	11 700	17 200	20 000	23 000	30 421	35 094

Resource: National Environment Agency, 1997-1998.

If, however, we take into consideration the numerous cars from the whole country that enter Tirana every day, the actual numbers are larger. Thus we can understand that urban traffic pollution predominates in Tirana. Preliminary studies show that in 1998 in Tirana cars released 5.7 t of gases and soot into the atmosphere. If every car in Tirana is driven for about 30 km/day they release a total of 9 459 t/year of CO_2 , 1 268 t /year of SO_2 and 1 038 t/year of NO_3. This is a big problem, especially in some streets like Deshmoret e Kombit, Durresi Street, Konferenca e Peze, and Muhamet Gjollesha. As can be seen from Table 2, in almost all streets the maximum exhaustion level lies between 12 and 13 p.m. During one day a total of 242 688 cars move within the city limits of Tirana, while within one year this sum amounts to 88 581 120 cars.

These cars are in mostly very bad conditions and at the same time they use fuel of low quality. In addition, our road network is not prepared to face this big number of vehicles. In order to bring this chaotic situation under control, some drastic measures are necessary: to discipline the use of fuel, to control the quality of fuel and vehicles, to limit the use of cars inside the inner city, etc.

TABLE 2. Consumption of fuel for cars and house heating in Tirana.

Fuel	Unit	1990	1991	1992	1993	1994	1995	1996	1997	2000
Firewood	m³	166 060	25 000	65 000	86 898	25 313	15 000	8 000	6 500	10 000
Coal	ton	10 897	1 420	1 200	1 293	50	0	0	0	0
Oil for Burners	ton	11 064	10 773	10 500	10 994	16 982	17 500	10 000	20 244	25 000
Liquid Gases	kg	0	0	0	182 495	164 816	570 000	90 000	1 323 020	7 000 000

Resource: National Environment Agency, 1997-1998.

Fuel and House Heating. After 1991, under the new economical and social conditions, a lot of changes in the use of fuels can be noted. Since there are not enough data available, we will only concentrate on Tirana.

As you can see from Table 2, there is a tendency to increase the consumption of liquid gases and oil, while the use of wood and coal is decreasing. Also the consumption of electric power increased 2 to 3 times. The quantity of coal burnt in 1997 is 20 244 t, which means that the quantity of CO_2 released by it amounts to 65 835 t. The total quantity of CO_2 predicted for 2000 is 104 280 t.

TABLE 3. Amount of NO_x in the air of Tirana in 1998.

City	Measuring Points	I	II	III	IV	V	VI	VII	VIII	IX	X
Tirana	Municipality	9.3	17.2	34.2	37.2	45.6	33.2	31.3	33.3	23.3	13
Durrës	Dalani Bridge	13.7	11.9	17.9	15.2	28.2	21.4	16.2	37.7	20.2	21.5

Resource: Public Health Institute.

From the above data it is obvious, that both the average and the maximum monthly values exceed the tolerable levels 3 to 4 times.

TABLE 4. Amounts of harmful gases in the air of Tirana in December 1997.

Harmful Gas	Min (ppm)	Max (ppm)	Monthly Average (ppm)	Tolerable Level (ppm)
SO_2	0	76	3.36	1.2
NO_2	0	48.5	5.21	2.4
CO	0.4	2768.1	194.27	900
O_3	3.7	41.4	23.56	5.4

Resource: Public Health Institution.

The high quantity of these gases in December is a result of the exaggerated use of fuel during this period for heating purposes, which is still favoured by specific meteorological conditions like temperature inversions, low pressure zones, etc.

In addition to chemical contamination, *air pollution by soot and dust* is very problematic. This kind of pollution is caused by the burning of wood and oil which makes up 11% of the total quantity of dust and soot in the city's atmosphere. Furthermore, it is also caused by intense urban and industrial activities, construction activities, large areas of bare soil and an increase of old vehicles, a reduction of green areas, a lack of hygiene in the cities, etc. As a result of all these factors, over some cities which are located in "dust and soot bowls" often smog clouds are formed.

TABLE 5. Amounts of harmful gases in the air of Tirana in January 1997.

	Min (ppm)	Max (ppm)	Monthly Average (ppm)	Tolerable Level (ppm)
SO_2	0.1	150.3	2.52	1.2
NO_2	1	53	3.34	2.4
CO	1.4	388.6	49.53	900
O_3	9.2	76.7	43.49	5.4

Resource: Public Health Institution.

TABLE 6. Solid particles in the air of Tirana with less then 10 microns, January 1997.

Measuring Points	Date	Duration (hours)	Concentration (mg/m³)	Average Concentration
	8-24/1/97	133.3	137.84	
	24-28/1/97	69	142.12	
Pediatric Hospital				139.98
	6-9/1/97	47.9	161.31	
	15-18/1/97	32.7	388.05	
Pediatric Hospital	20-28/1/97	59.8	284.82	299.5
	28-30/1/97	33.9	363.82	
	8-11/1/97	15.6	497.71	
	15-28/1/97	26.9	299.7	
Ambulance Nr.10	28/1-4/2/97	100.4	277.64	358.35

Resource : Public Health Institute.

TABLE 7. Solid particles in the air of Tirana with less then 10 microns, June 1997.

Measuring Points	Date	Duration (hours)	Concentration (mg/m³)	Average Concentration
	30/5-2/6/97	100	73.81	
	6/6- 9/6/97	67.9	115.13	
Pediatric hospital	13/6-16/6/97	67	204.93	131.29
	30/5- 2/6/97	68.8	76.91	
	6/6-9/6/97	72.8	142.51	
Municipal of Tirana				109.71
	30/5- 2/6/97	22.7	152.09	
	6/6-9/6/97	73.4	173.54	
Ambulance 10	13/6-16/6/97	43	212.96	179.3

Source : The Office of Air Quality.

Also the measurements made by HMI in Tirana and Elbasan concerning the content of dust in the air show that the values exceed the thresholds. The monthly limit is 350 mg/m²/day, and the annual value is 200 mg/m², while the measured values amount to

513 mg/m²/day in Tirana and to 533 mg/m²/day in Elbasan. Although high, these values do not exceed the critic value of 700 mg/m²/day.

2.3.2. *Pollution of Waters*

Communist Period. During the communist period the main pollution (about 25 million m³ of polluted waters a year) was due to industrial remnants like: nitrates, ammonia, hydrocarbures, cellulose, mercury, suspended organic and inorganic materials, etc.

Among the main pollutants of waters were:
- the metallurgic combine of Elbasan, which discharged up to 35 million m³ of polluted water per year into the Shkumbini River,
- the mechanic combine that polluted the Lana River

- The objects of chemical industry like:
- the sodium plant PWC in Vlora that damaged the shallow waters of the beaches
- the nitrogen plant in Fier that polluted the Gjanica River
- the plants for chemical and explosive substances in Tirana and Durrës
- the mines which contaminated with their mainly solid remnants the waters of the Fani, Mati, and Shkumbini Rivers etc.
- The petrol industry discharged 1.4 million m³ per year of contaminated waters, containing about 150 - 300 mg/l of petrol hydrocarbures
- the paper mills, which discharged up to 30 t/day of organic substances damaging the fauna
- the textile and alimentary combines etc.
- the solid and not treated liquid urban remnants.

Amongst the most polluted ones of the rivers one can mention: Shkumbini, Fani, Semani, Gjanica, Osum, Devolli, Lana, Kiri, Urake, and Drini River. In some of them the toxic substances and the continuous diminution of oxygen (until its full absence) have damaged the water flora and fauna, and have at the same time made these waters useless for irrigation. Best examples are the catchments of Shkumbini and Fani River in which before the beginning of the copper industry the biologists determined 18 families and 45 species of animals. In 1989 only 3 and 5 were found.

In some cases the ecological values of lakes are seriously endangered. From the most evident cases we pick out the following:
- The mouths of Kalasa and Bistrica River have been changed from draining into the Butrinti Lagoon to directly draining into the sea. This demolished the limnologic balance of this lagoon and spread H_2S. The change of the saline conditions caused a biologic catastrophe with mass extinctions of many water wildlife and organisms.
- The discharge of Devolli River into the Little Prespa Lake filled up its basin with solid materials, agricultural chemicals and urban remnants, that turned this very old lake into a polluted bog.
- The demolition of the ecological balance of the Ohrid Lake through human activities is more and more reducing the fish relic species. The major polluting factors of this lake are the agricultural chemicals, urban and industrial sewage

waters, in addition to the high concentration of people along certain parts of its banks.

- Lake Shkodra displays the same problems as Lake Ohrid. The discharge of the Lezha Drini River which flows through it changed the nutrition content. This phenomenon also affected its fauna.
- The carstic lake of Dumre and the glacial lakes of Lurë and Martanesh are exploited for irrigation without any limiting thresholds, thus lowering their levels under critical values, i.e. the biological minimum. This reduced their natural beauty and created big ecological stresses leading to the disappearance of many species.

Polluted waters of rivers have also contaminated the coastal waters, like in the Bay of Vlora, Drini etc. Contaminated materials reached the Albanian coastline also from other coasts of the Mediterranean Sea in the from of black hydrocarbure mud brought by the tides. Often this pollution came from the northern part of the Adriatic, originating from rivers which discharge there. All these factors have diminished the biological equilibrium of the Albanian coastal waters.

Transition Period. Besides the factors mentioned above for the air pollution, during this period we find several triggering factors for the water pollution. Among others foremost: the damage of the physical, chemical and bacteriologic water quality due to various reasons, where without any restriction subterranean water is used, channels are opened, industrial remnants are discharged, bottom materials of the river beds are exploited, etc. Sensible zones in this respect are the Korça Field, Fushëkuqe, Lushnjë and Vjosë. The zones with the highest pollution are the Shkodër - Lezhë Zone, Lezhë - Tirana, Elbasan, Lushnjë, Korçë, Berat and Fier. Serious problems are also noted in the carstic massifs which are well predisposed for pollution. For the coastal zones the salinisation of subterranean water has to be mentioned: in this context Velipojë, Shkodër, Shëngjin - Kune - Lezhë, Patok - Fushëkuqe, Shkozet - Durrës - Rodon, Hoxharë - Povelçe - Ura Mifolit - Fier, etc. are known.

Measurements have shown that during the period of 1997 to 1998 3350 t of contaminated solid materials and 12 450 t of liquid remnants have been deposited in rivers, lakes and the sea. The major polluting factors are still the petrol, gasoline, cement, leather, mechanic, ceramic, textile, wood, and paper industry.

The monitoring of the Albanian rivers has shown that most of them have a high water quality (from first to third level). At the same time, we can find some water bodies which are very polluted, like the Semani River or Gjanica with its liquid remnants of the petrol industry: $2.6 - 3.6$ mg/l of phenol remnants, while the legal maximum is at $0.02 - 0.05$ mg/l. This corresponds to a need of oxygen of $131 - 157$ mg/l, while the permitted level amounts to $8 - 12$ mg/l. In the Tirana River at Dunaveci, and in the drainage and irrigation channels like Roskovec-Hoxharë and Marinza the pollution is sometimes higher than common national and European norms.

The monitoring of lakes has shown that almost all are under the eutrophic level. Albanian lakes also contain only small quantities of organic phosphor and nitrogen. In the sea there is no oxygen glutting which, in some cases, is accompanied with high values of biological and biochemical oxygen consumption. This phenomenon is particularly noted north of Ishmi, in Durrës and in Vlora. The sectors of Durrës Beach,

Vlora and Pogradeci exceeded the allowed norms for chloroform excrements, while the streptoped excrements are within the allowed norm (except of the new Vlora Beach where there are high values). The waters of Dhërmiu, Borshi and Saranda Beach, are clean and within the norms defined by the European Union.

Today Albania is preparing the operational monitoring of its waters which will help to determine the environmental situation and to take more efficient measures for the improvement of the situation. Albania signed international conventions for well administration and the protection of springs, coastal waters, and the border water bodies. With the support of foreign organisations and donators projects for the improvement of the situation are under way.

2.3.3. *Pollution and Damage of Soil*

During the *communist period* near industrial and urban centres, along polluted rivers and in the agricultural zones where pesticides were used in big quantities, polluted soils have been created. This is due to the lack of "ecological culture" and technological control of agricultural products, causing a reduction of the quantity and quality of agricultural products in the zones with the polluted soil.

Among the big polluters were:
- the metallurgic combine of Elbasan with its discharge into the Shkumbini River which has been polluted with ammonia, phenol, cyanides, etc.
- the chemical enterprise of Durrës which released about 1 400 t bicarbonate remnants per year
- the paper mills in Lezha and Kavaja
- petrol industry, mines, colour metallurgy, etc.

Presently Albania has soils with "inherited" pollution. At the same time, although with less intensity, the soil is still being polluted by industrial activity (petrol, chromium, copper), by the urban remnants, etc.

Soil Damage by the Intensive Erosion. The erosion is 100 to 1 000 times bigger than in most other European countries. In average there is a total eroded area of 1 457 km²/year with an average eroded layer of 0.7 mm/year, the maximum value being 6 to 8 mm/year. As a consequence, many erosive forms like tarrents and hearths were created, even leading to badlands like in the Kërraba Hills, in the Dangëllia Highland, in Tomorrica, etc. where in some parts the ground is even desertifying. There exists a big number of natural and artificial reasons for this intensive soil degradation

Some of the *natural reasons* are:
- Over 60% of Albania's territory consists of lithified terrigenous material which is easily corroding.
- Over 25 % of the slopes with a steep inclination have a corrosion index of 1.0; those with an inclination of 15 ° - 20 ° have a corrosion index of 0.8 and those with 5 ° - 15 ° a corrosion index of 0.5 - 0.6.
- The Mediterranean climate is known to be very favourable for erosion, because the precipitation shows an average quantity of 600 - 3 100 mm/year, very irregular temporal distribution and high intensities. I.e. rains with 25 - 50 mm/hour intensity

amount to 47%, rains with 50 - 100 mm/hour intensity take about 24 %, and those under 25 mm/hour only 26%.

Light soils which prevail in Albania show 32 % more corrosion than the heavy ones.

Vegetation has only a limited protective role, because it covers just 51 % of the territory.

Some of the *anthropogenous reasons* are:
- the ploughing of new land on steep slopes and the clearing of shrubs or forests
- the inappropriate composition of crops (predomination of cereals, damage of vineyards and orchards, favoured locations for olive groves, etc)
- antiquated technologies in land cultivation
- soil pollution through the use of chemicals and the deposition of industrial and urban remnants, etc.
- antiquated technologies in irrigation
- intensive exploitation of the soil, etc.

The present political situation stimulates the soil damage because of several reasons:
- the temporary and permanent abandonment of the soils
- there is no periodic control of soil fertility and no law or any national institute in charge of soil protection
- the lack of a real capital law for land and no execution of the existing laws about it
- no national initiative for the fiscal stimulation to favour crop structures that foster soil protection from erosion.

Some recent aspects of erosion reduction are the following:
- Land privatisation is increasing the owners' interest to protect their soils.
- tendencies for changes in the crop structure in favour of fodder plants and orchards
- movement of the rural population into the cities
- Efforts in the direction of modern soil cultivation technologies and a decrease in the use of chemicals.

Marine erosion. In many parts of the Albanian coastline strong erosion can be observed. This also hampers the planning of wildlife reserves. Along the whole coast from Bunë to Vlorë erosion is noted at the Vlora Beach, in the previous-delta of Vjosa, in Seman, in the northern part of the Shkumbini River mouth, in the southern part of the Durrës Beach (Berg I Carpet to Gloom), in the Lalëz Bay, in the Patok Beach (with an intensity of 10 m/year), in the northern part of the Drini Bay, in the Buna Delta, etc.

This wide distribution and high intensity of maritime erosion is connected with many general and local factors, with the frequent changes of the river mouths and with anthropogenous factors, like the human intervention on rivers for hydrotechnical constructions (artificial lakes and dams, drainage and irrigation channels, etc.), the deviations of the river courses of Drini, Gjadër, Ishmi, etc., the disappearing of old rivers beds and coastal dunes; the exploitation of beach sands, the enormous gravel exploitation in the river beds, etc.

The entire coast from Vlora to Ftelis Bay (Greece) is *almost abrasive*, with an intensity of the marine erosion of 20 - 50 cm/year.

A solution for this environmental problem will require:
1. perennial coverage of the soil with vegetation
2. the regulation of the water regime with drainage as priority factor, not only with respect to agronomic problems but also for soil improvement
3. forest cultivation with fast-growing woods and native flora regeneration
4. modification of the plant coverage and the local streak of the plants
5. restoration of all hydrotechnical constructions that will stop the erosion, rinsing and sliding of the soil
6. controlled use of the manure
7. complex geological-geomorphological studies.

2.3.4. *Damage of Forests, Pastures and Biodiversity*

The forests and pastures have had the strongest environmental stress. During the *communist period* about 300 000 ha of bush woods and forests, or 32% of the forest surface and half of the bush land have been replaced by cultivated land. Even the new reforestation programme (about 8 - 10 ha/year) could not heal the damage that was caused.

During the *transition period* the forest damage continued, even at a faster pace than before. This is connected to the large illegal cuttings. Only in the 1998 2 450 ha were cut, almost 1 000 ha in the Lushnja District, 316 ha in Fier, and 187 ha in Kruja. This is combined with the overgrazing, with the use of the forests for illegal constructions, with contraband trade of lumber, and numerous cases of fires in forests (1997: 840 cases; 1998: 600 cases). The forests are also damaged by diseases like processonarie, oak ash and chestnut cancer, and by atmospheric pollution with acidified gases, etc. During the last ten years no new reforestation or rehabilitation of forestally degraded zones have been made. This is even more aggravating the situation of the forest in Albania. The only effort so far was the forestation of about 6 000 ha within a project financed by the World Bank. There are, however, some efforts for the protection of forests and bush wood going on. The governmental entities specialised in this are presently reorganising themselves.

The natural habitats have been significantly modified by: the dense population in some places and its dynamics by deforestation , overgrazing, fires, heavy use of wild plants, the increase of the cultivated land, the efforts for the creation of polifited brush woods (15 000 ha), the reclamation of many bogs and lagoons which caused the disappearance of faunal habitats, the state dictate for a crop predomination of cereals, the use of old technologies for soil cultivation and the intensive exploitation of the soil, the heavy use of chemicals etc. All this has damaged the biodiversity and threatened many plant species. In the whole Mediterranean and Submediterranean regions it is very difficult to find strictly natural ecosystems

Studies have shown that 3 mushroom species (Cortinarius laniger, Hygrocybe spadicea, and Cudonia circinany) and 2 plant associations have disappeared, 24 plant species and 41 plant associations are in danger to disappear, 30 species of plants and 30 plant associations are listed in the "Red List", 108 plant species and 43 plant associations are rare, 48.9 % of the endangered plant species and plant associations are very rare.

Studies have also shown that the situation of the fauna is even worse. So 7 species have disappeared, 6 of which are fowl species (Ciconia nigra, Anser erythropus, Branta ruficollis, Tadorna ferruginea, and Burhinus oedicnemus) and 1 mammipherous (Cervus elaphus); 45 species are in danger to disappear, 77 species are expected to enter the "Red List"; 258 species are rare and 10 in trouble. Animals such as Aquila clanga (globally threatened), Circaetus gallicus (regionally threatened), Accipiter gentilis gentilis (rare), tetrao urogallus (rare), and Bonasa bonasia (rare) have disappeared or are in danger to vanish.

As one can see, the situation of the biodiversity is rather serious and the tendency is quite negative. This will require urgent measures for the protection of the threatened wildlife.

The existing administrative structures in Albania are not able to detect and solve the numerous problems of biodiversity protection, to determine a strategy and to provide a legal basis for this protection. More successful have been NGOs which have carried out studies and proposed strategies and laws for biodiversity protection.

TABLE 8. Composition of Albania's surface cover from 1938 to 1996 (in units of 1000 ha).

Surface Cover	1938	1950	1990	1995	1996
Land (total)	2 875	2 875	2 875	2 875	2 875
Agriculture	292	391	704	702	701
%	10.1	14	24	24	24
Forest	1 385	1 282	4 045	1 052	1 026
%	48.2	45	36	36	36
Meadows and Brushwood	913.2	816	417	428	446
%	31.8	28	15	16	16
Other	285.0	386	709	693	701
%	9.1	13	25	24	24

Source: Team of Authors: Strategy and Action Plan for Biodiversity Protection, Tirana 1999.

For the *protection of biodiversity* the following is indispensable:
- the accomplishment of a general study about the new problems and situations created after 1990, after the break-down of socialist economy
- monitoring of the actual conditions of biodiversity
- active interaction with the farmers in order to obtain a consense between their interests and biodiversity protection
- to establish state authorities for the realisation of improvements of the crop structure with the aim to provide soil protection by vegetation during most of the year; to set up criteria for the use of chemicals, of new cultivation technologies and artificial irrigation
- The financing of the cultivation of degraded soils with medical plants or a modified natural herbage cover.
- To make farmers sensitive for ecological agriculture, implying modern soil cultivation technology and a lower use of chemicals.

3. Protected Landscapes

The establishment of protected zones started in 1940 when the hunting reserve of Kune Vain-Tale and the Tomorri National Park were proclaimed. In 1960 six national parks (Dajti, Thethi, Lure, Llogara, Divjake and Drenova) were created. Until 1970 15 hunting reserves were announced, containing lagoonal and forest zones. In 1975 their number increased to 25, and in 1985 the first nature monument was announced.

After 1992, on the basis of studies and concepts of IUNC and the approbation of an environmental law, the network of protected areas has been restructured. Presently the total area of protected landscapes amounts to 109 048 ha, i.e. 10.4% of the forest area or 3.7% of Albania's whole territory. According to the IUCN classification the following protected areas have been determined: strictly protected reserves (scientific reserve) with 14 500 ha, 12 national parks with 54 940 ha, nature monuments with 4 650 ha, 27 naturally managed reserves with 42 948 ha, 4 protected landscapes with 29 873 ha and 4 reserves with managed resources with 18 200 ha total.

Despite the present progress in the system of protected areas remains the solution of big practical problems and of the extension of their area, the improvement of the classification system, the review of the state of integration along with other types of ecosystems, in order to create representative functional zones for today and future. At the same time, the protected area management has to face all the problems of the transition period and the tendency for occupation of land for tourism and buildings. Thus, these areas have to be preserved according the IUCN regulations. Presently, however, illegal cuttings, pasturage and unlawful land occupation are going on in the protected areas. This phenomenon is most pronounced in the national parks of Lura, Valbona, Qafështamë, Bredhi i Hotovës, and Llogara. Many damages occur in the hunting reserves of Kune Vain, Rrushkull, Pishë Poro, Levan, Maliq, and Cangonj. In 1997 the ecological reserves of Karavasta were heavily affected, where apart from the burning of 8 ha of pine forest also an oasis for rare animals like wild cows, roebucks, etc. has been destroyed.

The entire coastal zone is mostly of high ecological value. It has been occupied and heavily damaged, by the buildings, which are constructed without any conditions in a rather chaotic way. If this is not controlled and stopped, these "wild settlements" will totally reduce the ecological value of the coastal zone.

4. Damage of Urban Environment

The uncontrolled and disorganised migration of the Albanian population has been accompanied by abusive construction, especially at the peripheries of the cities where there is the total lack of infrastructural services. This has caused an urban overpopulation with many social and environmental problems and is most evident in the Tirana - Durrësi - Elbasani - Fier Zone and in the Region Lezhë - Kurbin.

Tirana whose population has increased from 243 000 inhabitants in 1990 to about 700 000 in 1998, presents one of the environmental "hot spots". It combines a high concentration of people with insufficient urban infrastructure and has thus a strong influence on the environment. If the necessary measures will not be taken very soon the city development will be out of control causing a fast degradation of urban environment and, hence, the living quality. Presently, in Tirana the phenomena mentioned in the following sections can be detected:

4.1. UNLAWFUL CONSTRUCTION

First we want to mention "kiosk phenomenon": over 3 500 kiosks with serious environmental impacts. These reselling kiosks damage and occupy squares and main roads, public gardens and parks, green surfaces in general. Some years ago Tirana had a green surface of about 13 m^2/inhabitant. Despite other plans, however, in recent years the green surface has been significantly reduced due to damages, last not least caused by the increasing population movement into the capital. Today Tirana has a total green surface of about 2.3 million m². Five to six years ago there existed about 4 million m². Recently the surface per inhabitant decreased to 3.2 m²/inhabitant. The Grand Park of Tirana shows serious damages. For this park permission for the construction of hotels has been given to the Canadian firm Globex and to the Kuwait Alkaraf as well as for some private investigators for house construction.

It has to be noted, however, that some efforts improve the situation are on their way. E.g. there is an initiative for the destruction of illegal constructions, but again, the surfaces released from them are not green yet. They are covered by construction material and trash. Also, a greenness plan for Tirana is in preparation.

Illegal constructions have also damaged the infrastructure and sewage system. So, surface waters came into the sewage and the quantity of solid remnants increased very fast. This caused a hazardous health situation for the population with the risk of epidemic diseases. Especially at the periphery of the capital where we find a lot of "wild" constructions without any infrastructure there exists a very difficult situation.

4.2. INCREMENT OF URBAN RESIDUES

During 1998 about 520 000 t of urban residues were produced in Albania, 44 % of which came from the five cities Tirana, Durrës, Vlora, Shkodra, and Elbasan. This quantity corresponds to 255 kg/inhabitant per year. Within the last year, these residues increased by 8 - 10 %, which shows the urgent necessity to take measures on a sound scientific basis which presently lacks totally. As a consequence, the health risk increased, soils and subterranean waters are contaminated, and the aesthetics of the urban landscapes is ruined.

Scientific approaches to the problem of urban residues are made within the scope of the EU-funded LIFE Program. Its intention is to prevent the creation of residues from the very beginning. They have to be recycled, and the bulky solid remnants to be collected.

5. Industrial Remnants

During the communist period enormous quantities of solid materials accumulated like mining industry remains, enriched and melted minerals, cement, TEC, etc. E.g. the copper industry generated over 389 500 t of void rocks and toxic industrial remains, from the chromium industry result 9 million m³, from the coal mine in the Tirana Basin 450 000 t of void rock material.

In 1998 the quantity of solid industrial remnants reached a total of 415 000 t, where the chromium and cooper industry took the first place. Presently, the treatment of new and old industrial remnants is one of the most pressing environmental problems.

In addition to these remains, there exist also highly dangerous toxic substances in big quantity. At the first place one has to mention 8 000 m³ of arsenic in the bunkers of the Fieri Nitrogen Enterprise which are licking, thus creating a big danger for the surrounding of this plant and for the whole Adriatic Sea. The 1 000 t of pesticides which are beyond the using data and have remained unused, but must be cleaned up ver quickly also belong to this category. In order to solve this problem a strategy based on a sound legal basis has to be developed.

6. Concluding Remarks

Only the most troubling environmental problems of Albania could be treated in this paper. In summary, we can state that as a consequence of the long-lasting pollution and of the "old" and the "new" degradation about 20 % of Albania were polluted and will remain like that. These 20 % represent Albania's "hot spots". In Albania is also evidence of a considerable desertification process which is very advanced in some places.

At the same time we want to stress that the major part of Albania is clean and displays some areas of natural beauty and high ecological values. The approbation of legal bases, the efforts by governmental institutions and environmental NGOs, the support of specialised international institutions, and the efforts made to materialise environmental education, represent positive signs for the environmental protection in Albania.

7. Bibliography

1. National Environmental Agency, *Report on the Environmental Situation during 1995-1996*.
2. National Environmental Agency, *Report on the Environmental Situation during 1996-1997*.
3. Akademia e Shkencave, Flora e Shqipërisë; Scientific Academy, *Flora of Albania*, First part, 1989.
4. Demiri M, Gjeografia e bimëve, *Geography of Plants*, Tirana, 1973.
5. Demiri, M. Flora eskursioniste e Shqipërisë, *Excursion Flora of Albania*, Tirana 1985.
6. Haxhiu I, Zvarrinikët në Shqipëri, *The Reptile of Albania*, Tirana 1978.
7. Haxhiu I. Amfibet në Shqipëri, *Amphibians of Albania*, Buletini i Shkencave Natyrore, 3, 1985.
8. Harta e botës shtazore, *Map of the Fauna of Albania*, Tirana, 1978.
9. Instituti Hidrometeorologjik, Klima e Shqipërisë, Hydrometeorological Institute, *Climate of Albania*, Tirana 1975.
10. Instituti hidrometeorologjik, Pasuritë ujore dhe Klimatike të Shqipërisë, Hydrometeorological Institute, *Water and Climate of Albania*, Tirana 1980.
11. Instituti Hidrometeorologjik, Atlasi Klimatik i Shqipërisë, Hydrometeorological Institute, *Climatic Atlas of Albania*, Tirana 1990.
12. Instituti Hidrometeorologjik, Hidrologjia e Shqipërisë, Hydrometeorological Institute, *Hydrology of Albania*, Tirana 1985.

30

13. Instituti i Studimeve dhe i Projektimeve Gjeologo-Minerare, Harta Gjeologjike e Shqipërisë, Institute of Geologic Studies and Projects, *Geological Map of Albania*, Tirana 1983.

14. Kolektiv autorësh, Strategjia dhe plani i veprimit në mbrojtje të biodiversitetit, Team of authors, *Strategy and Action Plan in Protection of Biodiversity*, Tirana 1999.

15. Kolektiv autoresh, Libri i kuq për bimët, Team of Authors, Red Book of Plants, Tirana, 1998

16. Kolektiv autorësh, Raporti i gjendjes së mjedisit në Shqipëri, Team of Authors, *Report on the Environmental Situation in Albania*, Tirana, 1993.

17. Kofina M. Gjendja e Mjedisit për qytetin e Tiranës, *The Environmental Situation of Tirana*, Tirana, 1997.

18. Mitrushi I. Drurët dhe Shkurret e Shqipërisë, *Trees and Brushwood of Albania*, Tirana, 1954.

19. Qiriazi P. Gjeografia Fizike e Shqipërisë, *Physical Geography of Albania*, Tirana, 1998.

20. Qiriazi P. *Terrenet e degraduara të rrethit të Tiranës, Degraded Terrains of Tirana District*, Studime gjeografike, Nr. 4, 1991.

21. Qiriazi P. Samimi E., *Issues of the Geographical Environment of Albania*, Tirana, 1995.

22. Qiriazi P. Sala S., *Format erosivo - denuduese në pellgun ujëmbledhës të Shkumbinit dhe disa vleresime të intesitetit të erosionit të këtej pellgu ujëmbledhës; Erosive Forms in the Catchment of the Shkumbin River and Some Assesments of Erosion Intensity in this Basin*, Studime Gjeografike, Nr. 9, Tirana, 1996.

23. Raport mbi veprimtarinë e bashkisë së Tiranës për vitet 1992 - 1996 dhe prognoza për vitin 2000; *Rapport of the Municipality of Tirana for 1992-1996 and Prognosis for 2000*.

24. Studimi i Ndërrmarjes së Gjelbërimit, *Study of Greenness Enterprise*, Tirana, 1996.

25. Xinxo Z. Erozioni në Shqipëri, *Erosion in Albania*, Tirana, 1989.

AVAILABILITY OF CURRENT SPACEBORNE EARTH OBSERVATION DATA

WOLFGANG BAETZ
Gesellschaft für Angewandte Fernerkundung mbH (GAF)
Arnulfstraße 197
80634 München
Germany

Abstract

Over the past 30 to 40 years Earth Observation (EO) has undergone a tremendous development, not only with respect to the manifold applications feasible today, but also with respect to the technologies utilised to handle space borne information. In the beginning it was the synoptic information, which triggered scientists and especially geologists when analysing the first space borne photographs. Snapshots presenting large areas of the earth's surface revealed features and interrelationships of earth scientific phenomena never seen before.

In the seventieth the US government initiated and implemented the so-called LANDSAT program, which is still in operation today, and which consisted over the years of a series of EO satellite systems. For the first time these automatic-operating satellites provided a constant and complex information flow from space.

The success of the Landsat program has spawned many similar earth resources satellites by several other nations as well as private industries. Presently more than 20 earth observation systems are providing data on a routine basis for operational applications in various fields, e.g. cartography (map updating, topographic and thematic base mapping), land cover/-use assessment, and monitoring environmental conditions on land and at sea. Different orbit configurations are used, and satellite sensors can view the Earth in vertical, side, or stereo modes.

1. General Characterisation of Satellite Data

Compared to ground observations remotely sensed satellite data show important advantages. Satellite images provide a synoptic and repetitive overview of the Earth's surface. In addition, the near global, repetitive collection of the data using satellite sensors is cheaper than collecting the same type and quantity of information using conventional methods, e.g. ground survey, aerial photography.

The information content of the space borne imagery is limited by the data characteristics in terms of spectral, temporal and spatial resolution.

M.F. Buchroithner (ed.),
Remote Sensing for Environmental Data in Albania: A Stragegy for Integrated Management, 31–40.
© 2000 *Kluwer Academic Publishers. Printed in the Netherlands.*

Spectral resolution stands for data recorded simultaneously and separately in several portions of the electromagnetic spectrum utilising atmospheric windows. The acquired ultraviolet, visible infrared, and microwave energy coming from the Earth's surface or atmosphere contain a wealth of information about material composition and physical conditions.

Temporal resolution stands for the repetition rate. It leads to repeated images of the same regions, taken at regular intervals over periods of hours, days or years provide data bases for recognising and measuring environmental changes.

Spatial resolution describes the smallest unit to be identifiable on an image. The spatial resolution is described per Picture Element (Pixel). It may range from 1 m (very high resolution) to 1 km or several kilometres per pixel (very low resolution). Weather satellites for example, which aim at a high frequent coverage are typically characterised by very low resolution.

2. The Main Operational EO Systems

Landsat started out with 80 m resolution systems. Today multi-spectral cameras typically operate in a range between 30 m and 20 m such as the Landsat Thematic Mapper (TM), the French Spot system with 20 m and the Indian IRS-1C and 1D EO satellites with 23 m.

Concerning high-resolution panchromatic data the market is currently dominated by 10 m Spot data and 5m IRS data. This segment is actually being completed by the Landsat 7 system with the sensor Enhanced Thematic Mapper (ETM) providing 15 m spatial resolution, launched early 1999.

Nowadays, a lot is demanded of remote sensing systems. They should provide the user with reliable, up-to-date information sufficient for a broad range of diverse application fields in different scale levels. The IRS system for example integrates three different cameras. They simultaneously provide multi-spectral and panchromatic information in different levels of resolution, such being suitable for manifold applications in environmental monitoring, land management, mapping, etc. The Wide Field Sensor (WIFS) acquires data with 2 spectral bands, covering 800 km x 800 km at 180m resolution, a multi-spectral camera (LISS-III) with 4 bands is covering 140 km x 140 km at 23 m and a panchromatic camera (PAN) with 1 band covers 70 km x 70 km at 5.8 m resolution.

Figure 1 shows a natural colour composite of IRS-1C-Pan/LISS data with 5m resolution, covering the international airport of Tirana, Albania.

Figure 1. Natural colour composite of IRS-1C-Pan/LISS data with 5m resolution, showing the airport of Tirana (Albania) Please see appendix for image in colour.
Copyright: ANTRIX/SII/euromap 1999

3. A New Generation of Very High Resolution Data

Extraordinary technological developments in computer and communication applications as well as the user requirements actually drive the development of new satellite systems with improved overall performance, particularly with respect to spatial resolution and fast availability of the data. Nevertheless, due to limited transmission rates and processing capacities a trade-off between scene size (synopsis), number of spectral bands and spatial resolution has to be made.

This new generation of very high resolution data is provided by the IKONOS system, the world's first and only commercial high resolution imaging satellite, which

34

was launched September 24, 1999. The IKONOS satellite sensor simultaneously collects one meter resolution black-and-white (panchromatic) images and four meter resolution colour (four band multi-spectral) images at a scene size of 11 km x 11 km. Designed to take digital images of the Earth from an orbit of 680 km, the satellite camera can distinguish objects on the Earth's surface as small as one meter square in size. At one meter resolution one is able to see cars and trucks, roads, pipelines, individual trees, houses, large equipment, boats and ships, airplanes, and other objects at least one meter in size.

Figure 2 displays one of the first IKONOS images in colour with 1m resolution. The processing was done fusing the 1m resolution black/white data with 4m colour data. The sample shows a part of Beijing city, China.

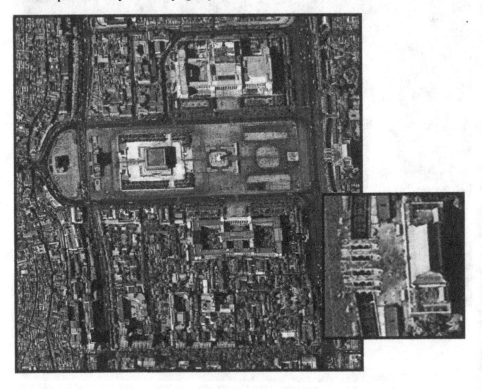

Figure 2. 1m resolution IKONOS data showing a part of Beijing (China). Please see appendix for image in colour. Copyright: SI 1999, GAF 1999

IKONOS imagery will be made available for sale to customers in the beginning of 2000.

Table 1 provides an overview of specifications of the main currently available satellite sensors. Figure 3 shows the size of the standard scenes combined with the spatial resolution of the single systems.

TABLE 1. Overview of main current operational satellite systems.

Platform	Spectral range	# of bands	Resolution (m)	Image frame / Swath width (km)	Repeat cycle (days)	Launch
Optical Digital Scanner Systems						
NOOA-K (15)	VIS, NIR SWIR, MWIR TIR	2 2 2	1000	3000	12 hours	May 1998
SPOT-VEG (4)	VIS, NIR SWIR	3 1	1000	2250	26	March 1998
Resurs-01	VIS, NIR SWIR TIR	4 1 1	225 810 810	714	16	July 1998
IRS-1C/D-WiFS	VIS, NIR	2	188	810	24	1996/1997
Landsat- MSS (5)	VIS, NIR	4	80	180	16	1984
Landsat-TM/ETM (5, 7)	VIS, NIR SWIR TIR PAN	4 2 1 1	30 30 120, 60 15	180	16	1984/1999
SPOT 1, 2, 4	VIS, NIR SWIR PAN	3 1 1	20 20 10	60	26	1986/1994/ 1998
IRS-1C/D	VIS, NIR SWIR PAN	3 1 1	24 72 5,8	140 140 70	24	1996/1997
IKONOS-2	VIS, NIR PAN	4 1	4 1	11	11	Sept. 1999
Russian Analogous Camera Systems						
KFA-1000	VIS, NIR	2	5 - 10	120 x 120	Up to several years	Single missions
KFA-3000	PAN	1	2	21 x 21	Up to several years	Single missions
KVR-1000	PAN	1	2 - 3	40 x 40	Up to several years	Single missions
MK-4	VIS, NIR	6	8 - 10	170 x 170	Up to several years	Single missions
KATE 200	VIS, NIR	3	20	225 x 225	Up to several years	Single missions
TK-350	PAN	1	10	200 x 300	Up to several years	Single missions
Active Microwave Systems						
ERS-2	C-BND-VV	1	25	100	35	1995
RADARSAT	C-BND-HH	1	6 - 28	50 - 150	24	1995

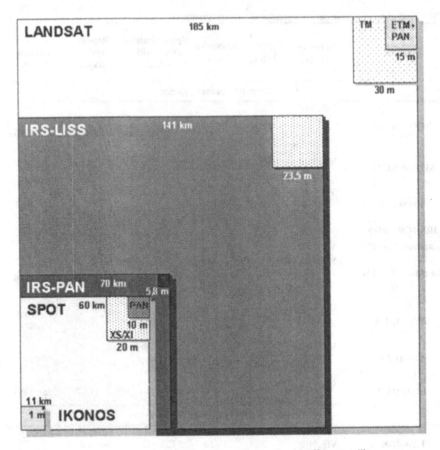

Figure 3. Coverage comparison and spatial resolution of main different satellite sensors.

4. Categories of Spatial Resolution

According the overview given in Table 1, the optical satellite imagery can be grouped as follows:

Low resolution satellite data – 250 m to some km spatial resolution by multi-spectral sensors, e.g. NOAA, SPOT-Vegetation and Resurs

Medium resolution satellite data – 80 m to 180 m spatial resolution by multi-spectral sensors, e.g. Landsat MSS, Resurs, IRS-1C/D-WiFS

High resolution satellite data – 5 m to 30 m spatial resolution by panchromatic or multi-spectral sensors or analogue camera systems, e.g. Landsat TM, SPOT-PAN/XS/XI, IRS-1C/D-PAN/LISS, KFA-1000

Very high resolution satellite data – 1 m to 4 m spatial resolution by panchromatic or multi-spectral sensors or analogue camera systems. For the private field of RS applications, until September 1999, the data of the Russian camera systems (e.g. KFA-3000, KVR-1000) are the only available.With IKONOS 2 digital data with 1 m in pan, and 4 m in multi-spectral mode will become available beginning of the year 2000.While spectral and temporal resolution refer to the thematic information extraction, the spatial resolution is directly related to the coverage provided and the maximum possible mapping scale. See Table 2: Satellite data - mapping scales.

TABLE 2. Satellite data - mapping scales.

Resolution	Sensor	Mapping scale
Low (< 250m)	SPOT-Vegetation NOAA	< 1 : 1 000 000
Medium (80m – 180m)	Landsat MSS RESURS IRS-1C/D-WiFS	1 : 200 000 – 1 : 1 000 000
High (5m – 30m)	TM/ETM SPOT-XS/XI/Pan IRS-1C/D-LISS/Pan ERS Radarsat KFA-1000 MK-4 KATE 200 TK-350	1 : 50 000 – 1 : 200 000
Very high (> 4m)	IKONOS KFA-3000 KVR-1000	1 : 5 000 – 1 : 25 000

5. Data Availability and Distribution

In addition to the national or international governmental organisations operating space programs and systems, such as the US NASA, the Indian Space Research Organisation ISRO, or the European Space Agency ESA, the commercial leaders in providing satellite data to a broad satellite remote sensing user community – researchers and consultancy industry - are represented by the following organisations dominating the market:

Spot Image, France
Space Imaging, US
Antrix, India
Radarsat, Canada
Sovinformsputnik, Russia.

These organisations provide a broad array of information products and services, off-the-shelf imagery from different sources for all information needs, and acquire data on request to fulfil specific user requirements.

38

Space Imaging (SI) as well as the other organisations offers not only data collected by their own systems. Based on exclusive or non-exclusive distribution agreements, they also market data from competitors. For example, SI (US) running IKONOS, got the world-wide rights from ANTRIX, the commercial branch of the Indian Space Research Organisation (ISRO), to market the Indian satellite data of the IRS-1C/D systems. To cover a customer segment as broad as possible, SI additionally markets data of the systems Landsat Thematic Mapper (TM), Japanese Earth Remote Sensing Satellite (JERS), European Radar Satellite (ERS), Radarsat, and also aerial photographs. Data receiving and archiving is organised by a world-wide network of ground stations run by the system operators itself or from organisations, which are licensed to receive and archive the data.

Figure 4 illustrates the SI access to a network of international ground stations. The data distribution is organised through an international network of representatives or affiliates of the various organisations operating the satellite systems and ground stations.

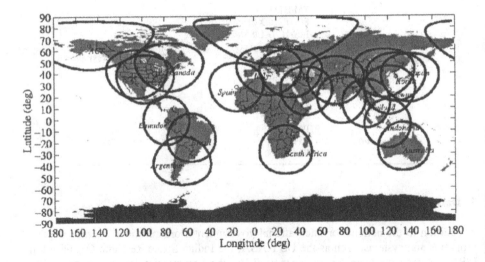

Figure 4. Space Imaging network of international ground stations. Please see appendix for image in colour.

In general these representatives, located all over the world, are specialised in value adding or consultancy services in the fields of remote sensing and GIS. These companies have re-seller agreements with at least one, but very often with more or all of the above global leaders of satellite system operators.

In Europe, apart from the system operators such as Spot Image (France) that directly sell the data, typical re-sellers are the Italian company Eurimage or the German company GAF. Eurimage markets data from Landsat, ERS, IRS, JERS, and Russian imagery through a network of international representatives. GAF for example has signed distributor agreements with all commercially most important data providers (SI, Antrix, Radarsat, Spot Image, Sovinformsputnik). Moreover, together with its daughter company Euromap (based near Berlin), GAF/Euromap was the first private European company operating in close co-operation with the German research centre DLR its own ground station for the reception and distribution of the Indian data exclusively for Europe.

6. Costs for Satellite Data

Satellite data, which is far cheaper compared to data collected by alternative systems or methods (aerial photography, ground survey), is sold to the end users according to internationally valid pricelists provided by the system operators. The pricing depends on the scene sizes (standard full scenes, sub-scenes, or in parts with a minimum sales unit in km^2) and the different pre-processing and processing levels (e.g. system corrected up to precision geometric corrected). Table 3 indicates costs per data set of standard full scenes, respectively per km^2.

TABLE 3. Cost comparison of main data sets (standard radiometric and system corrected data).

Sensor	# Bands	Resolution (m)	Coverage (km)	Costs/Scene (Euro)	Costs/km^2 (Euro)
Landsat TM (5)	7	30 (120)	180 x 180	3 500	0.11
Landsat ETM-(7)	7	15, 30 (60)	180 x 180	1 500	0.05
IRS-1C/D-LISS	4	23	140 x 140	2 700	0.14
SPOT XI (4)	4	20	60 x 60	2 600	0.72
SPOT XS (1,2)	3	20	60 x 60	2 100	0.58
IKONOS XS	4	4	11 x 11	2 178 US$ (World 3 509 US$)	EU 18 US$ (World 29 US$)
SPOT PAN	1	10	60 x 60	2 600	0.72
IRS-1C/D-PAN	1	5,8	70 x 70	2 500	0.51
KVR-1000	1	2-3	40 x 40	24 000 US$	15 US$
KFA-3000	1	2	21 x 21	11 025 US$	25 US$
IKONOS PAN	1	1	11 x 11	2 178 US$ (World 3 509 US$)	EU 18 US$ (World 29 US$)
IKONOS PAN/XS	3	1	11 x 11	2 904 US$ (World 4 598 US$)	EU 24 US$ (World 38 US$)

Up to now, because of the huge data volumes, the delivery is mainly done by mail or courier services, and the data is stored according to international standard formats on CD-Rom, Exabyte or other media. Using modern communication technologies, "on-line delivery", especially if small data sets are concerned, is becoming more and more important and will be the standard in the near future.

7. Conclusions

Mission objectives of the above mentioned EO satellite systems are to continuously collect information about our environment. Taking into consideration the long period of time most of these satellites have recorded and transmitted EO data, powerful data archiving, and information retrieving systems are indispensable in order to fully take advantage of the data.

Internet based image browsing systems have initiated a development which allow to access data and to make available recently acquired and also historical information to a multidisciplinary and global user community within minutes. Still remains, what we call data exploitation for application related analysis. It is obvious, that only a small fraction of once recorded information can be analysed - nevertheless many applications still suffer from lacking information due to the fact, that very often, data taken from a certain region at the right point of time might not be available. This situation will gradually improve with the implementation of more EO systems in the coming decade.

8. Internet Sources describing Data and Data Suppliers / References

http://www.spotimage.frhttp://www.rsi.ca/
http://202.54.32.164/ (NRSA India)http://www.sovinformsputnik.com/http://www.esrin.esa.it/
http://www.euromap.de
http://www.eurimage.it
http://www.gaf.de
http://www.spaceimaging.com/

9. General Guide Describing Sensors, Programmes and Application

http://www.ceo.org/

MAPPING FROM SPACE

GOTTFRIED KONECNY
Institut fuer Photogrammentrie und Ingenieurvermessung
University of Hannover
Nienburger Str. 1
30167 Hannover, Germany

1. Why Mapping from Satellites?

The motivation for mapping from space is given by the fact that past conventional mapping methods have not been able to provide adequate mapping coverages at the required scales, except for priority areas.

According to statistical surveys carried out by the U.N. Secretariate (World Cartography 1993) the following world mapping status has been achieved:

1.1. GLOBAL MAPPING

In form of the Digital Chart of the World DCW compiled by the U.S. Military Agency NIMA from existing maps of different countries of the world a digital data set at the scale 1:1 000 000 is now publicly available for the cost of a few hundred dollars. However, due to the different geodetic reference systems and due to the different object classifications used this data set is non-homogenized with geometric errors in the 10 km range.

NIMA attempts to compile a better global reference data set at the scale 1:250 000 by the end of 1999, in which the geometric errors are reduced by the existence of available GPS observations. The homogenization efforts are carried out in different portions of the globe, for example by the UN-FAO Africover project for parts of Africa.

These basic data sets will provide a valuable asset for other international coordination efforts such as "Global Mapping" carried out by an ISPRS Working Group in cooperation with national mapping agencies.

1.2. NATIONAL MAPPING

National mapping provides the basis for regional planning within the framework of sustainable development. Its aim is to provide a land area map coverage at scales 1:50 000 or 1:25 000 by the efforts of national mapping agencies.

Sofar (according to the U.N. reports) a 67 % global coverage has been achieved for the scale 1:50 000 (for developing nations), and a 33 % global coverage for the scale 1:25 000 (for developed nations) (see Figure 1).

41

M.F. Buchroithner (ed.),
Remote Sensing for Environmental Data in Albania: A Stragegy for Integrated Management, 41–58.
© *2000 Kluwer Academic Publishers. Printed in the Netherlands.*

42

However, the surveys of current map updating progress indicate, that the worldwide average age of the existing 1:50 000 maps is 45 years, and of the existing 1:25 000 maps 20 years. While the situation is somewhat better in Europe, where update rates range from 7 to 15 years in continents such as Africa and Latin America the update rates are more than 50 years (see Figure 2).

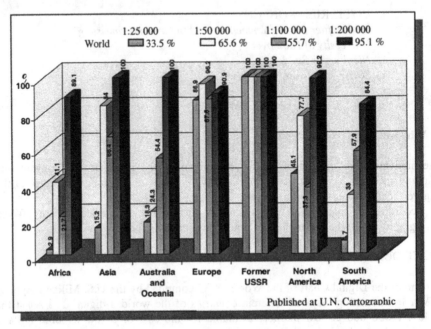

Figure 1. Status of World Mapping

1.3. LOCAL MAPPING

No reliable data are available on the large scale mapping coverages at scales 1:10 000 or larger. In general, a world wide lack of data has been noted.

Only few countries, such as Germany and Great Britain have large scale mapping systems at scales ranging from 1:5 000 to 1:1200, which are maintained for local and urban planning and the maintenance of a property cadastre.

The worldwide lack of large scale mapping data is directly related to the high mapping cost by conventional technologies such as ground surveys and aerial mapping. These cost factors have limited such mapping efforts to priority areas.

2. Existing Operational Cartographic Satellite Systems

A number of existing operational satellite systems can be used for mapping. Their suitability is governed by their resolution, their coverage (swath) and their repeatability (depending on cloud penetration).

2.1. METEOROLOGICAL SATELLITES

Geostationary meteorological satellites, such as Meteosat by ESA, GOES by NOAA, GMS by NASDA and Insat by ISRO, permit imaging every half hour at 5 km resolution.

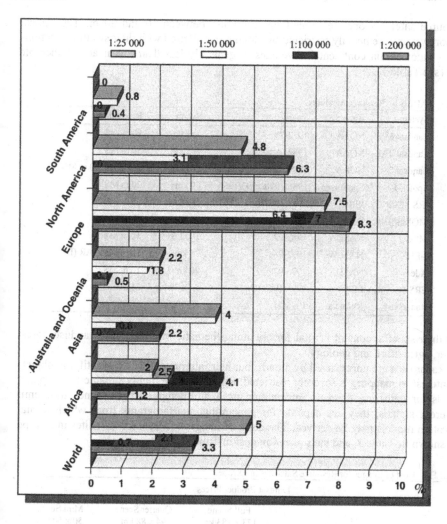

Figure 2. Status of World Map Updating.

The widely used NOAA satellites permit imaging an area 2 to 4 times per day at 1 km resolution, reaching cloudfree area coverages every 10 to 15 days. They are ideally suited for global and regional meteorological issues and for the monitoring of global natural and planted vegetation. Their thematic data may be ideally merged with global base data sets at scales 1:1 000 000 or 1:250 000. The satellite data are easily available (at reduced resolution even through the Internet) at negligible cost.

2.2. RESOURCE SATELLITES

Resource satellites operate at medium resolution between 10 and 30 m. The optical sensor systems are heavily restricted by cloud cover. Paired with their swath restrictions they cannot obtain continental coverages in generally less than a year at considerable cost (see Table 1).

TABLE 1. Resource Satellites.

System	Agency	Years	Swath	Resolution
Landsat MS	NOAA	1972-78	185 km	80 m MS
Landsat TM	NOAA	1982	185 km	30 m MS
Landsat 7	NOAA	1999	185 km	15 m pan, 30 m MS
Spot 1-4	Spotimage	1986/1990/1993/1998-	60 km	20 m MS
IRS 1C/D	ISRO	1995/1997	142 km	23 m MS
MOMS 02P	DLR	(1993)1996-	78 km	18 m MS
MOS	NASDA	1982-92	100 km	50 m MS
J-ERS 1	NASDA	1992-98	25 km	18 m MS, SAR (L)
Adeos	NASDA	1996-97	80 km	16 m MS
ERS ½	ESA	1991/1995-99	50 x 70 km	25 m
Radarsat 1	Radarsat	1995	50-500 km	10-100 m

But their spectral content is vital for obtaining thematic information on landuse, forest cover, agriculture and geology.

Radar data are unrestricted by clouds, but heir interpretability for identifying objects of interest to mapping is severely restricted. Radar data are highly suitable for mapping floods for gathering thematic information on a multitemporal basis. Since they emit coherent signals, they are capable for generating interferograms from which digital elevation models may be derived. The costs of images of Resource Satellites in Europe are shown in Table 2, and the square km cost in Table 3.

TABLE 2. Cost of Images for Resource Satellites in Europe.

Landsat Product Prices		
	Full Scene 173 x 183 km	Quarter Scene 94 x 88 km	Mini Scene 50 x 50 km
7 channels	4000 $	2200 $	1800 $
1 channel	2000 $		
7 channels geocoded via orbit data	5500 $	3000 $	2600 $
7 channels geocoded via control points	6000 $	3200 $	2900 $
7 channels geocoded via control and DEM	6800 $	3500 $	3200 $

Spot Product Prices						
	Standard 60 x 60 km	Radiometric Correction Level 1A	Rad. & Simple Geom. Correction Level 1B	Orbit Corrected Level 2A	Grd. Control Level 2 B	Ortho-photo
XS	1400 $	2300 $	2300 $	3000 $	3200 $	3300 $
PAN	1700 $	2800 $	2800 $	3500 $	4000 $	4200 $

IRS-1C Data				
	70 x 70		23 x 23 km	
	Orbit Corrected	Control Corrected	Orbit Corrected	Control Corrected
PAN	3100 $	3300 $	900 $	1000 $
LISS III	3300 $	-	1600 $	-
WIFS 806 x 806 km	-	-	900 $	-

ERS-1/2 Data	
Full Scene	450 $
Control Geocoded Image	1500 $
Terrain Geocoded Image	2000 $
Multitemporal scene (3 images)	2200 $
Interferometry Set	3000 $

JERS-1 Data	
Precision Image	1200 $
Geocoded Image	1500 $

Radarsat Data	
Standard	3400 $
Orbit Corrected	4000 $
Orbit Corrected	5000 $

Russian Scanned Photography KVR 1000	
Imagery	28 $/km^2
Geocoded Imagery	60 $/km^2

NOAA-AVHRR	
Image	150 $

TABLE 3. Costs of Images per km^2 for Resource Satellites in Europe.

LANDSAT Satellite Data Costs per km^2			
	173 x 183 km	94 x 88 km	50 x 50 km
Landsat TM, 7 channels	0.13 $	0.26 $	0.70 $
Landsat TM, rectified to orbit data	0.17 $	0.38 $	1.04 $
Landsat TM, rectified to control	0.19 $	0.40 $	1.13 $
Landsat TM, rectified to control & DEM	0.20 %	0.44 $	1.31 $
Landsat MSS, 4 channels	0.01 $	-	-
Landsat TM, rectified to orbit data	0.05 $	-	-
Landsat TM, rectified to control	0.06 $	-	-
Landsat TM, rectified to control & DEM	0.07 $	-	-

SPOT Satellite Data Costs per km^2			
	60 x 60 km Image	Orbit Data corrected	Control corrected	Control & DEM corrected
SPOT-XS	0.36 $	0.68 $	0.89 $	0.94 $
SPOT-P	0.47 $	0.87 $	1.08 $	1.18 $

IRS-C Satellite Data Costs per km^2		
	70 x 70 km	23 x 23 km
IRS-C Pan	0.63 $	1.83 $
Map Geocoded	0.68 $	1.95 $
LISS II 4 channels	0.68 $	3.96 $
Map Geocoded	0.74 $	
WIFS 806 x 806 km	0.001 $	

ERS-1 Satellite Data Costs per km^2	
ERS-1 100 x 100 km	0.05 $
Geocoded	0.17 $
DEM Geocoded	0.28 $

RADARSAT Satellite data Costs per km^2	
RADARSAT 50 x 50 km	1.92 $
Geocoded	2.16 $
DEM Geocoded	2.50 $

2.3. CARTOGRAPHIC SATELLITES

The recent U.N. Open Skies Agreement has opened the way to commercial high resolution systems, which will provide superior mapping capabilities.

This path has been initiated by the off track stereocapabilities by SPOT 1 to 4 and IRS 1, as well as by the longtrack stereo system of MOMS 02P and the higher resolutions of 6 m achieved by MOMS 02-P, IRS 1/C and the Russian mapping film cameras KFA 1000 and KVR 1000 operated from the MIR platform (see Table 4).

TABLE 4. Cartographic Satellites.

System	Agency	Years	Swath	Resolution	Stereo
Spot 1-4	Spotimage	1986/1990/1998	60 km	10 m pan	off track
IRS 1C/D	ISRO	1995/1997	70 km	5.8 m pan	off track
KFA 1000	RKK	.	66-105 km	5 m	no stereo
KVR 1000	RKK	.	22 km	2 m	no stereo
MOMS 02-P	DLR	1996-	37 km	6 m	along track
Adeos	NASDA	1996-97	80 km	8 m pan	off track

1999 the era of 1 m resolution systems, which has been announced since 1996 is expected to become reality (see Table 5).Iconos 2 was successfully launched in September 1999.

TABLE 5. Future Cartographic Satellites.

System	Agency	Expected launch	Swath	Resolution	Stereo
Ikonos 2	Space Imaging	Sept 1999	11.3 km	0.82 m	along track
Quick Bird	Earth Watch	1999	22 km	0.82 m	along track
Orbview 3	Orbimage	1999	8 km	1 m	along track
Eros B	West Indian Space	1999	13.5 km	1.3 m	along track
Spot 5	Spotimage	2001	60 km	3 m	along track

3. Suitability of Satellite Images for Mapping

Mapping requirements can be expressed by 3 parameters:

- the position accuracy
- the elevation accuracy
- the object detectability.

3.1. POSITION ACCURACY

Map accuracy standards usually define the position accuracy as ± 0.2 mm in the required map scale, as given in Table 6.

TABLE 6. Position Accuracy Requirements.

Map scale	Standard deviation of positioning
1:5 000	± 1 m
1:10 000	± 2 m
1:25 000	± 5 m
1:50 000	± 10 m
1:250 000	± 50 m

3.2. ELEVATION ACCURACY

Conventionally map accuracy standards defined the permissible contour interval, which was chosen as a function of the slope of the terrain. Flood planes required a contour interval of 1 m or 2 m, hilly or mountainous areas were satisfied with 20 m or 50 m contours. Contours are to describe the elevation of a point at the 90 % confidence level.

In aerial photogrammetric mapping procedures the contour interval chosen was usually 5 times the standard deviation for determining the height of a point. Nowadays contours are interpolated from D.E.M. measurements carried out in grid-fashion, which permit to automatically interpolate contours at 3 times the standard deviation of a point determination.

The mapping requirements related to a scale can in general be expressed as in Table 7.

TABLE 7. Heighting Requirements.

Map scale	Standard deviation of point height	Contour interval
1:5 000	± 0.3 to ± 0.6 m	1 to 2 m
1:10 000	± 1.5 m	5 m
1:25 000	± 3 m	10 m
1:50 000	± 6 m	20 m

3.3. DETECTABILITY OF OBJECTS

Detectability of objects depends on the desired map content. For the objects to be mapped the values given in Table 8 should be maintained.

TABLE 8. Detectability of Objects.

Object type	Pixel size
houses	2 m
paths	2 m
toads	5 m
creeks	5 m

3.4. POSITIONING REQUIREMENTS AND CURRENT SATELLITE SYSTEMS

The positioning requirements of existing satellite systems may all be met, if a sufficient number of control points is available for an image. 4 to 8 such well distributed control points will generally suffice for the existing digital sensors to which the Landsat, Spot, IRS 1 or Moms images are fitted by geometrical resampling using polynomials.

Film images of the KVR 1000 have higher distortions, so that more than 8 control points are needed to eliminate the film distortions to ± 2 m for the scale 1:10 000.

3.5. HEIGHTING REQUIREMENTS AND CURRENT SATELLITE SYSTEMS

Photogrammetric height accuracy not only depends on image scale (or pixel size), but also on the elevation parallax determinable from 2 overlapping images from different exposure stations, characterized by the height to base ratio.

Table 9 lists the height to base ratios achieved by different satellite systems.

TABLE 9. Height to base ratios for satellite missions.

	h/b	elevation accuracy
Spot	1/1	± 5 m
Spot	2/1	± 10 m
IRS 1C/D	1/1	± 3 m
IRS 1C/D	2/1	± 6 m
MOMS 2P	1.5/1	± 8 m

This means, that the achievable height accuracy for satellite systems is marginal; for off-track stereo systems, such as Spot and IRS 1 it is moreover hampered by having to obtain 2 cloudfree scenes on subsequent dates with possible long time intervals, during which changes in reflectivities are likely.

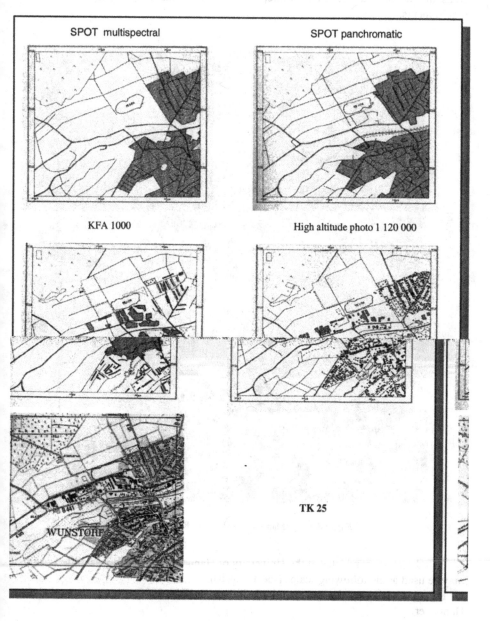

Figure 3

50

3.6. DETECTIVITY AND CURRENT SATELLITE SYSTEMS

Figure 3 demonstrates that the chosen detailed content of a German map 1:25 000 depicting buildings cannot be achieved with pixel sizes of less than 2 m. If settlement areas are to be mapped as a whole, then 10 m pixels may suffice for reduced requirements.

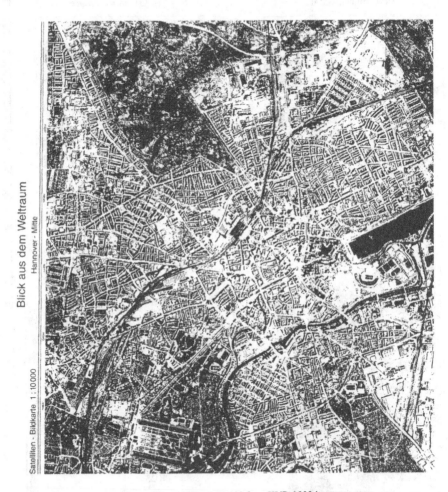

Figure 4. Image Map 1:10 000 from KVR-1000 imagery

Based on tests conducted at the University of Hannover the existing satellite systems may be used at the following scales (see Table 10).

Figure 4 shows an image map 1:10 000 from KVR-1000 imagery compiled for Hannover.

TABLE 10. Suitability for Mapping.

system	pixel size	Image map	line map
Landsat	30 m	1:100 000	1:250 000
Spot	10 m	1:50 000	1:100 000
IRS1 C/D	6 m	1:25 000	1:50 000
MOMS 02P	6 m	1:25 000	1:50 000
KVR 1000	2 m	1:10 000	1:20 000
Spot 5	3 m	1:15 000	1:25 000
Ikonos	1 m	1:5 000	1:10 000

3.7. RADAR INTERFEROMETRY

Due to the marginal performances of optical satellite stereo systems for height determinations radar interferometry is of considerable interest for D.E.M. determinations.
At the University of Hannover a test was carried out whether a D.E.M. could be derived from the ERS 1/2 tandem mission, which would meet mapping requirements. A test area of 10 x 10 km^2 was used for which an accurate DEM with ± 1 m accuracy was available.

The results shown in Figure 5, 6 and 7 show that accuracies of ± 5 m could be reached in flat unforested areas, but that deviations of up to 100 m were obtained in areas of radar forshortening, radar shadows, and in forested areas.

4. Cost Comparisons

4.1. COST OF SATELLITE DATA VERSUS COST OF AERIAL PHOTOGRAPHY

The cost of satellite imagery has always been a bone of contention except for governmental and sponsored users.

While the worldwide distribution of Landsat 1 & 2 MSS imagery in the 1970's and early 1980's helped to spread satellite remote sensing technology and its wide application al around the globe it was the Landsat Commercialization Act of 1985 in the USA, which brought about a change toward commercialization. Spot in 1986 was built up as a system with a commercial component. High licensing fees for the reception of Spot and Landsat have forced reception stations to shut down because of insufficient image sales. Several governments being in the position to operate satellite programs subsidized the space segment, but not the transformation of data into information, the so-called value-added business.

In this situation the traditional mapping organizations and the mapping industry were reluctant to enter an uncertain market with a product technology and its marketing yet to be developed.

Despite this fact some successful business ventures utilizing e.g. available Russian imagery have been initiated. At the advent of commercial high resolution satellites it is a very opportune time to review the advantages of satellite mapping with respect to cost.

The costs of imagery per km^2 shown in fig 5 show that Landsat TM data at current prices range from 0.13 $ to 1.31 $/km^2; those of Spot from 0.36 to 1.18 $/km^2; those of IRS1 from 0.63 $ to 3.96 $/km^2, those of ERS from 0.05 to 0.28 $/km^2, and those of

52

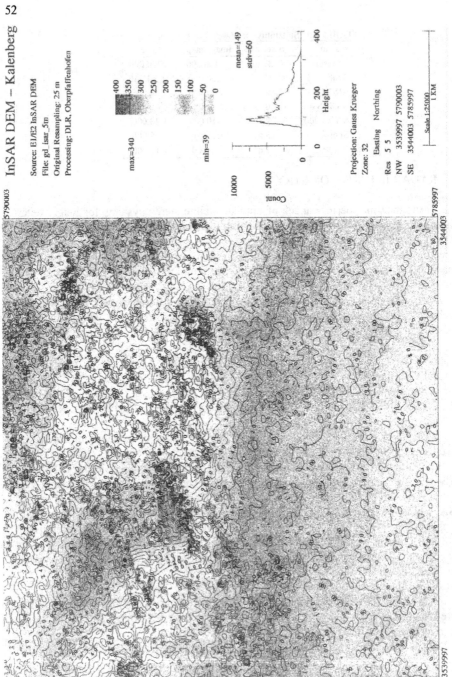

InSAR DEM – Kalenberg

Source: E1/E2 InSAR DEM
File: gd_isar_5m
Original Resampling: 25 m
Processing: DLR, Oberpfaffenhofen

max=340

400
350
300
250
200
150
100
50
0

min=39

mean=149
stdv=60

Projection: Gauss Krueger
Zone: 32
        Easting   Northing
Res   5   5
NW   3539997 5790003
SE   3544003 5785997

Scale 1:25000   1 KM

Figure 5. Interferogram.

53

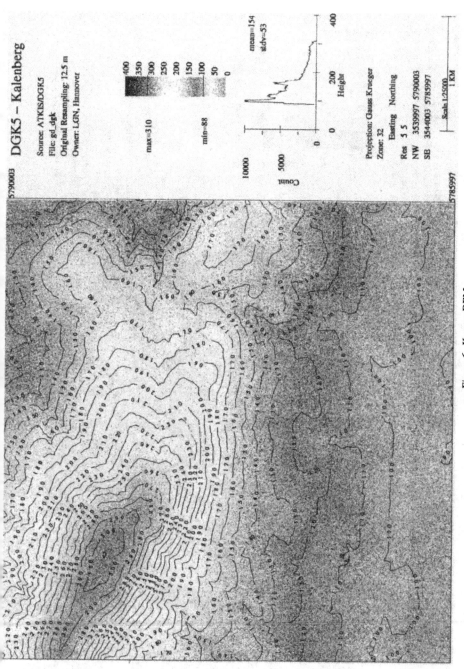

Figure 6. Known DEM.

54

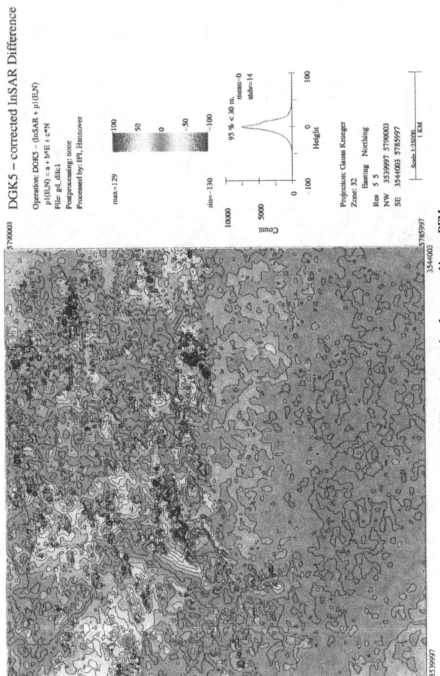

DGK5 – corrected InSAR Difference

Operation: DGK5 – (InSAR + p1(E,N)
  p1(E,N) = a + b*E + c*N
File: gd_dilc1
Postprocessing: none
Processed by: IPI, Hannover

max=129

100
50
0
–50
–100

min=–130

95 % < 30 m.
mean=0
stdv=14

Count
10000
5000
0

–100   0   100
Height

Projection: Gauss Krueger
Zone: 32
        Easting   Northing
Res   5   5
NW   3539997   5790003
SE   3544003   5783997

Scale 1:25000
                1 KM

5790003   3544003

5783997

3539997

Figure 7. Differences between interferogram and known DEM.

Radarsat from 1.92 to 2.50 $/km². The variation is due to the size of data set ordered and the processing level offered, but it also reflects the rather non-transparent differences between subsidy and commercial interest.

While no standard values for aerial photography are available throughout the world due to differences in weather restrictions, in access to planes and in size of the project, some basic scale dependent averages can nevertheless be quoted in Table 11.

It is obvious that satellite mages are in general much cheaper.

TABLE 11. Cost of Aerial Photography.

aerial photography scale	cost per km²	resolution
1:80 000	4 $/km²	1.2 m
1:30 000	8 $/km²	50 cm
1:6 000	16 $/km²	10 cm

4.2. COST OF MAPPING PER STEREO MODEL

The cost of mapping is scale dependent. For a particular scale for which each photo covers a number of km² it is composed of

- the cost per photo
- the control required for the restitution
- the measurement of the digital terrain model
- the line mapping of objects
- verification and addition of ground checked data

The restitution of a single stereo model composed of two overlapping photographs requires a minimum of 3 to 6 control points, which help to find the orientation of the image pair with reference to the geodetic reference system on the ground. A control point can today be established with differential GPS satellite positioning at the rate of 50 $/point. The process of aerial triangulation in which models are joined by measurement of corresponding image points at the rate of 30 $/model helps to reduce the control point requirement if a whole block of photos is evaluated.

The measurement of a digital terrain model can be accomplished in about 5 hrs of operator time.

Digital photogrammetry, which requires scanning of the images at the rate of 15 $ per image, permits to perform aerial triangulation and DEM generation in a more automated way. This reduces the time required for the operation, but not the cost.

The bulk of the effort consists in the operator executed line mapping of object which can vary from 10 hrs per model in rural areas to 100 hrs per model in urban areas. Thus the cost per model restitution can range from 630 $ to 4230 $, if an operator/machine hour is assessed as 40 $/hour.

The only cost effective alternative possible by digital photogrammetry is to produce a differentially rectified image, the so-called "orthophoto". In this product, which can be generated at 120 $ per model, the measured D.E.M. is used for the computer controlled automatic resampling of images into a geocoded form corresponding to the geometric accuracy of a map. Thus an orthophoto can be generated for 385 $ per model, which is 2 times to 11 times cheaper than the preparation of a line map.

The extraction of a line map is therefore the expensive and time consuming portion of the restitution process depending on the object content.

The digital orthophoto technology from aerial photos is equivalent to the provision of geocoded satellite data. While the value added creation from both aerial and satellite images is a tedious process, the geocoded image creation is affordable for both types of imagery.

This may be used to advantage when old maps are to be updated. Provided they possess the required position accuracy (if not, they need to be fitted to new control), they can be raster scanned or manually digitized for better feature separation and then superimposed with the digital orthophoto. In that case only the changed portions of the features need to be removed or added by on-screen digitization, which is a significantly smaller effort connected with less time and cost.

Digital photogrammetry with aerial or satellite images is therefore the only viable solution for the global map update problem.

4.3. COST OF MAPPING PER AREA

Satellite mapping becomes of ultimate advantage, when efforts and costs are compared on the basis of the area covered.

An example may demonstrate this: a metropolitan area of 1000 km^2 is to be covered by aerial photography. For a maximum flying height of 10 km an image scale of 1:33000 can be achieved with a normal angle camera. This imagery can be digitized into 50 cm pixels. An image covers 7.6 x 7.6 km, the base between photos in a strop is 3.0 km and the distance between strips is 5.3 km. This gives a neat stereo model area of 16 km^2. 100 km^2 can thus be photographed by 63 images. The restitution cost of these images into orthophotos is thus about 25 000 $.

The Russian KVR-1000 imagery from an altitude of 200 km is available in 2 m pixels. An image covers 26 x 26 km or 676 km^2. 2 to 4 images can therefore cover the metropolitan area at a restitution cost of 1600 $. The cost advantage is therefore a factor of 12 while meeting the requirements for 1:10 000 image mapping.

For this very reason satellite image mapping utilizing the expected new satellite products has a chance to overcome the global, regional and local bottlenecks in map production and map updating. The parameters of these new sensors are contained in Table 12.

The image costs presently quoted for such systems are listed in Table 13. They are at least one order of magnitude less than the cost for value added products.

TABLE 13. Image Costs.

System	Cost/km^2
KVR 1000	15-30 $/km^2
Ikonos 2	30-100 $/km^2

TABLE 12. Commercial Earth Observation Satellites.

Systems	Earth Watch "Quick Bird"		Orbital Sciences "Orb View 3"		Space Imagery "Ikonos"		West Indian Space Ltd. "EROS"		Earth Watch "Early Bird"		Resource 21 "Resource 21"		GEROS		Kodak "Cibsat"		
Enterprises	Ball Hitachi Telespazio MDA		Orbital Sciences		Lockheed Martin E-Systems Mitsubishi		Israeli Aircraft Ind. Core Soft-ware Techn.		Ball Hitachi Telespazio MDA		Boeing Farmland GDE ITD		Geophys. & Env. Res. Corp. Space Vest				
Launch	1. 2000 2. 2001		2000		1. failed 2. Sep 1999		1. failed 2. 2000		failed		1. 2000 (2) 2. 2001 (2)		1. 2000 (2) 2. 2001(2) 3. 2002 (2)		cancelled		
Mode	Pan	MS	Pan	MS	Pan	MS	Pan		Pan	MS		MS	Pan	MS	Pan	MS	hyper-spectral
Quantization	11 bit	11x4 bit	8 bit	8 bit	11 bit	11 bit	10 bit		8 bit	8 x 3 bit		12 bit			11 bit	11 bit	
Resolution	0.82 m	3.28 m	1 & 2 m	4 m	0.82 m	4 m	1.3 m		3.2 m	15 m		10 m / 20 m / 100 m		10 m			
Channels	1	4	1	4	1	4	1		1	3		4 / 2 / 1			1	5	60
Swath	22 km		8 km		11 km		13.5 km		6 km	30 km		205 km			112 km		
Pointing in track	± 30°		± 50°		± 45°		± 45°		± 30°			± 30°	-		2 convergent sensors		
Pointing cross track	± 30°		± 50°		± 45°		± 45°		± 28°			± 40°	-				
Sensor position	GPS		GPS		GPS		GPS		GPS			GPS	GPS		GPS		
Sensor attitude	star trackers		2 star trackers		3 star trackers		-		1 star tracker			star trackers	star trackers		2 star trackers		
Expected accuracy with GCP's (horiz / vert)	2 m	2 m	7.5 m	3.3 m	2 m	3 m	6 m	4 m	6 m	4 m	5 m abs	1 m rel.	3 m	3 m	5 m	3 m	
Without GCP's (horiz / vert)	23 m	17 m	12 m	8 m	12 m	8 m	800 m		150 m			30 m	25 m				

58

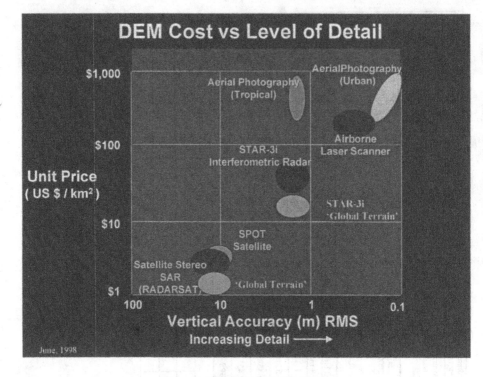

Figure 8. DEM Costs vs. Level of Detail.

In Figure 8 a scheme is presented (from B. Mercer), which compares D.E.M. costs for different achievable vertical accuracies commensurate with cost.

While aerial photography is expensive and other airborne sensors are intermediate in price, satellite sensors are cost effective.

While aerial photography an laser scanning still yields the highest accuracy, moderate vertical accuracies are still achievable via satellite technology.

WWW INFORMATION SERVICES FOR EARTH OBSERVATION AND ENVIRONMENTAL INFORMATION

NINA D. COSTA, MICHEL MILLOT, CLIVE BEST, BERNDT ECKHARD
Strategy and Systems for Space Applications Unit
Space Applications Institute
European Commission
TP 261
Joint Research Centre
I-21020 Ispra (VA)
Italy

Abstract

INFEO – Information on Earth Observation – is the CEO (Centre for Earth Observation) Project's on-line information system. INFEO represents a single access point for information on the availability of Earth observation (EO) related data products (i.e. metadata describing what exists) and it improves accessibility (stating how to get the data, sometimes on-line) to EO data and information. INFEO can be found at http://infeo.ceo.org/ .

The CEO Project is funded by the European Commission (EC) to develop and promote the use of Earth observation (EO) data from space. INFEO represents the CEO's initiative to make it easier for anyone to find, and gain access to EO and related data, information and services. INFEO replaces the CEO's first generation on-line information service for EO, the European Wide Service Exchange (EWSE).

INFEO allows users to query, in parallel, EO related catalogues throughout the world, ranging from large inventories of space data to smaller geo-referenced data sets. Any query submitted in INFEO is automatically submitted to all selected on-line catalogues. It thus provides users with a common access point for local or remote data and information, and a shared way of finding information.

INFEO also allows providers of EO data, information and services to advertise their products at a one-stop shop, thereby reaching a large number of customers through a single 'outlet'. Any holders of EO data or information can link their catalogues to this system so that they are included in any search.

This ideal of a one-stop shop for EO information is made possible due to the development of a common language (CIP - Catalogue Interoperability Protocol), which allows the querying of different remote data catalogues in a simple and consistent manner.

M.F. Buchroithner (ed.),
Remote Sensing for Environmental Data in Albania: A Stragegy for Integrated Management, 59–63.
© 2000 *Kluwer Academic Publishers. Printed in the Netherlands.*

1. European Wide Service Exchange (EWSE)

Despite the wealth of Earth observation data, information and services, potential customers were often unaware of their existence, since there was no single information system which is capable of searching and finding these services.

In September 1995, the CEO released for public use the first version of a system to provide the exchange of, and accessibility to, metadata and information about Earth observation, data and services. This system was called the European Wide Service Exchange (EWSE), and it represented the first Internet-based virtual marketplace for EO data and services. It allowed service providers to advertise their data, information and services, and customers to search for them.

A search for information on the EWSE could be carried out in a number of different ways: by free text, by keyword or by location (on the map of the globe). Organisations or individual users, registered with the EWSE, could also be found via alphabetical lists or via the Tradeshow, where organisations are arranged by thematic area. Throughout its lifetime, the EWSE was continually improved and updated. It was a huge success as reflected by its statistics; before it was phased out, it had registrations for over 3700 users, 970 organisations and 665 EO products.

Between 1996 and 1999, the CEO developed a more advanced system called Information on Earth Observation (INFEO) and in September 1999 the EWSE was replaced by this newer system. In addition to many of the features of EWSE, INFEO offers users the possibility to search for images (i.e. data rather than just metadata) in remote catalogues located world-wide. Most of information held within EWSE has been transferred to INFEO.

2. Catalogue Interoperability Protocol (CIP)

The catalogue search functionality of INFEO is due to its adoption of a communication protocol called Catalogue Interoperability Protocol (CIP). It should be noted that CIP, developed in an international partnership through CEOS (the Committee on Earth Observation Satellites) under the leadership of CEO, has now been formally adopted as the international catalogue interoperability protocol.

CIP is an application profile of an earlier communication protocol (z39.50) aimed at georeferenced datasets and, in particular, space datasets. Basically it defines the semantics of a search query.

CIP has also been promoted in the wider geo-spatial field, with an adaptation of CIP being proposed to the OpenGIS consortium as its official interoperability protocol. The OpenGIS consortium is an international body seeking to put "standards" in place for the GIS world. Again this is being undertaken under the CEOS umbrella.

CIP was selected for INFEO since it allows customers to search for data without knowing in which catalogue they might be located, or how a particular catalogue is structured and therefore how to query it.

3. INFEO

3.1. SEARCH OF REMOTE CATALOGUES

A search for data in INFEO can simply be based on a combination of geographic, temporal or thematic criteria. The user can, for example, search for all data over Ispra on 31 December 1998 or all data that show volcanoes. He/she does not have to specify the satellite or sensor, or know in which catalogue the relevant data is held.

The following satellite data catalogues are currently connected to and therefore searchable via INFEO:

- ARIS (Agricultural and Regional Information System) at the JRC
- ISIS (Intelligent Satellite Information System) at DLR
- EUROMAP at DLR
- MARF at EUMETSAT
- EURIMAGE
- SPOT
- HEOC (Hatoyama Earth Observation Centre) at NASDA

A facility was developed to interface with the IMS protocol compatible catalogues, which provides access to a further 800 datasets including all the American DAACs (Data Active Archive Centres), as well as to the Canada Centre for Remote Sensing.

The CEO aims to continually connect new space catalogues to the system in order to offer as comprehensive a service as possible.

In addition to all these satellite or space catalogues, INFEO users can also search a selection of relevant non-space catalogues from the same single user interface. These include the land cover catalogues from ITE (Institute for Terrestrial Ecology), UK and MDC (Swedish Environmental Data Centre), Sweden. Access is also offered to a map catalogue at IGN (Institut Geographique National) in France, an atmospheric catalogue at BADC (British Atmospheric Data Centre).

3.2. EUROPEAN ENVIRONMENT INFORMATION SERVICES (EEIS) PROJECT

In order to improve data interoperability between the two large user communities of Environment and Earth observation, the project 'European Environmental Information Services' was launched in early 1998.

Through this joint project undertaken by CEO and the European Environment Agency (EEA), the gap between the environment and EO communities has been bridged by linking INFEO to the EEA's Catalogue of Data Sources (CDS).

The main challenge of this project was to use the Catalogue Interoperability Protocol (CIP) to interconnect two radically different search systems (INFEO and WebCDS). Although CIP was originally intended for EO users, it has been expanded to cover non-space data as well.

EEIS will provide unique access to both types of information within an homogenous user interface and without reducing the performance of any of two services. Each user community will be able to access the service via their 'own' search systems: Environmental user will access via the locator system 'WebCDS+' (http://www.mu.niedersachsen.de/cds/eeis/) extended with new features to explore the data of remote sensing; Earth Observation users will access via the INFEO search service (http://infeo.ceo.org).

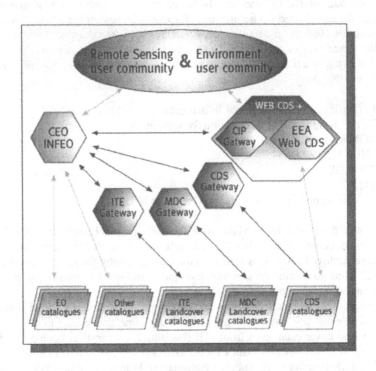

Figure 1. A schematic diagram explaining the EEIS project.

The EEIS service is new in that it will interconnect two existing search systems for related user communities. An Environmental user will be thus be able to easily find earth observation data (e.g. satellite images) for their research and reports, and vice versa. It will be launched early in 2000. For the latest information on this project please see the EEIS homepage at http://eeis.ceo.sai.jrc.it/ .

4. Conclusions

INFEO was developed in close cooperation with international partners to ensure coherence of approach, and to minimise development overlaps. This is mainly undertaken via the Committee on Earth Observation Satellites (CEOS) Working Group on Information Systems and Services (WGISS). Participation in CEOS WGISS has ensured that the INFEO development is coherent with other developments globally.

Through the launch of INFEO, the CEO is closer to realising it's ideal of offering a one-stop shop for EO data and related information.

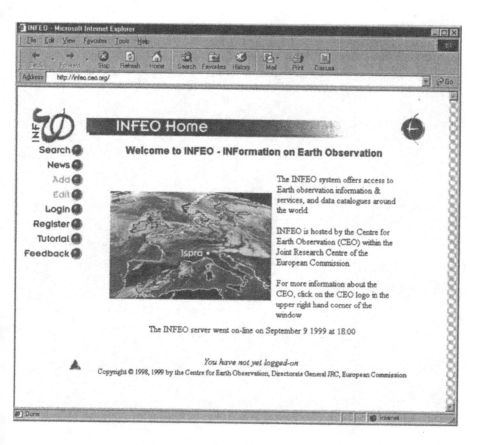

Figure 2. Screen shot of the INFEO home page. Please see appendix for image in colour.

5. References

1. CEOS WGISS PTT (1997) *Catalogue Interoperability Protocol (CIP) Specification – Release B Issue 2.2*, CEOS.
2. CEOS WGISS PTT (1998) *Catalogue Interoperability Protocol (CIP) Specification – Release B Issue 2.4*, CEOS.
3. Rolker, C., Kramer, R. and Kazakos, W. (1999) Interoperability among Earth Observation and General Environmental Data Catalogues via CIP, *Proceedings of the Earth Observation & Geo-Spatial Web and Internet Workshop '99*.

MAP MAKING WITH REMOTE SENSING DATA

THIERRY TOUTIN
Canada Centre for Remote Sensing
588 Booth Street,
Ottawa, Ontario, Canada, K1A 0Y7

Abstract

Map making with remote sensing data requires geometric and radiometric processing methods (monoscopic and stereoscopic) adapted to the nature and characteristics of the data in order to extract the best cartographic and topographic information. For the monoscopic method, different geometric and radiometric processing techniques are compared and evaluated, quantitatively and qualitatively with their impact on the resulting composite images, using panchromatic SPOT and airborne SAR images. The techniques that take into account the nature of the data give better results, with greater integrity: a subpixel geometric accuracy with high-quality composite images, which are sharp and precise and containing well-defined cartographic elements and data that are easy to interpret and closer to physical reality. The stereoscopic method still is the most common method used by the mapping, photogrammetry and remote sensing communities to extract three-dimensional information. It is successfully applied either to images in the visible spectrum or radar images to generate digital elevation model with an accuracy of tens of metres depending of the data source.

1. Introduction

Throughout history, humans have tried to represent what they saw and understood through images. Everything from cave walls, to canvases, to computer screens have been used to express perception of our surroundings. Maps have been one way to show the relationship between humans and their environment. Towns, roads, rivers, mountains, valleys, and where the land meets the sea, have been drawn in an organised fashion for centuries. Mapmakers have always sought ways in which to represent both the location and the three dimensional shape of land.

Not so long ago, a hill top view was the largest vista from which to observe nature's workings. Discoveries in optics, photography and flight have allowed us to see the Earth as never before. Advanced methods in computing and signal processing technologies have

M.F. Buchroithner (ed.),
Remote Sensing for Environmental Data in Albania: A Stragegy for Integrated Management, 65–87.

enabled us to increase our ability to visualise and perceive the Earth's surface. Today, Earth observation satellites orbit our planet collecting data needed to produce images which allow us to monitor, understand and plan the use of our world's resources. However, specific processing methods have to be performed on the satellite images to extract information before making maps.

Two conventional methods can be considered to extract information from remote sensing data (Figure 1):

1. The monoscopic method which uses one image and an existing digital elevation model (DEM) to generate an ortho-image from which only planimetric features with their 2-D map coordinates (XY) can be extracted; and

2. the stereoscopic method which uses two images to generate a "virtual" stereo-model from which planimetric and altimetric features with their 3-D map coordinates (XY and/or Z) can be extracted

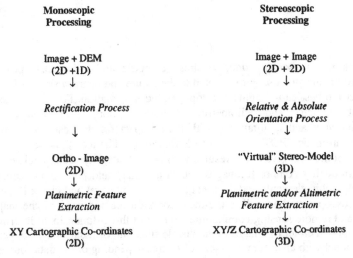

Figure 1. Description of the monoscopic and stereoscopic processing methods for 2D or 3D cartographic feature extraction.

In the first method, the DEM has to be produced from any method (contour lines digitising, stereoscopy, interferometry, etc.) with some errors. These errors will propagate through the rectification process and the planimetric features extraction. Furthermore, resampling during the rectification process degrades not only the image geometry and radiometry, but also the image interpretability.

In the second method, the brain can generate the perception of depth with two images from same or different sensors. The stereoscopic fusion of multi-sensor images then provides a virtual three-dimensional model of the terrain surface, and the stereo plotting enables the extraction of cartographic features directly in the map reference system. Conversely to the first method, the planimetric accuracy of feature positioning is not affected by any elevation error. Furthermore, since the stereo-extraction is done directly on the raw images, no re-sampling degrades the image radiometry, geometry and interpretability.

The main objective of this paper is to present the basic processing steps of these two conventional methods, which are necessary to generate maps from remote sensing data. Comparisons and performances of different techniques, tools and softwares for the processing steps are presented. Finally, some examples and results of map making with different satellite images from these two methods are showed.

2. Monoscopic Processing

The monoscopic processing of satellite images can be based on the concept of "geocoded images", to define value-added products [1]. Photogrammetrists, however, prefer the term "ortho-image" in referring to a unit of geocoded data. To integrate different satellite images under this concept, each raw image must be separately converted to an ortho-image so that each component ortho-image of the data set is registered pixel by pixel and the different radiometries can then be combined [2].

The composite images are products resulting from the integration of different images. Their creation requires two distinct processing steps to ensure that those elements, which are spatially and spectrally separable in the original images, are also separable in the composite images:

- geometric processing to ensure that each pixel in the ortho-images corresponds to the same ground element;
- radiometric processing to merge the information from each image in a common image, such that the best spectral information from each image is preserved.

There are many references in the literature, which combine and/or compare the data in the visible and microwave spectra. Early works by [3, 4, 5, 6] and many others dealt mainly with the integration of Landsat and Seasat data, although other geocoded data were also used [7].

The technique most commonly used is image-to-image registration with a previously geocoded reference image. This registration uses polynomial or spline functions with many tie points between the images. However, these authors generally report the difficulty of finding such tie points between the images, because they are imaged differently by sensors, which have highly variable geometries and responses to illumination. Errors resulting from this method are of a few pixels, which then generate errors in the radiometric merging of the various ortho-images. The effect is even more significant in mountainous terrain. It is not the best appropriate technique to generate maps.

Consequently, to demonstrate the interest of geometric and radiometric processing techniques suited to the nature and characteristics of the different images, and to measure the impact of the processing tools, it is a requisite:

- to compare different geometric and radiometric processing tools; and
- to quantitatively and qualitatively evaluate the impact of different processing tools on the resulting composite image.

Two techniques of geometric correction are compared: the polynomial functions generally used and a rigorous photogrammetric method developed at the Canada Centre for Remote Sensing (CCRS) [8]. The latter technique allows the integration of DEM into the correction for a better accuracy. Four techniques of merging the radiometric information from the resulting ortho-images are evaluated: red-green-blue, principal components, intensity-hue-saturation and high-pass filter. To enable the evaluation of the two geometric correction techniques significant and to be extrapolated, different images (VIR and SAR; spaceborne and airborne) are used:

- a SPOT-P raw image (level 1A) acquired June 20th, 1987 at a highly tilted viewing angle (+29.3°) and the ephemeris and attitude data related to this image;
- four airborne synthetic aperture radar (SAR) images (north-south flight direction) acquired September 11th, 1990 by the CCRS radar (C-HH, narrow mode, angle of 45° to 76°, ground distance, 4096 pixels by 10 000 lines, pixel spacing of 4.0 by 4.31 m) [9].

Since the width of a SAR image swath is approximately 16 km, two adjacent images were taken pointing east and two others pointing west to create two radiometrically different SAR mosaics over the test area (26 by 40 km). The SPOT-P image has a grey-scale dynamic range of 17 to 60. No radiometric processing was done, except linear stretching on 8 bits. The SAR images were processed in real time in the aircraft and were encoded on 8 bits. No radiometric processing was done of these images.

2.1. GEOMETRIC PROCESSING

While it is known that polynomial functions are not suitable for accurately correcting airborne or space images, many users still apply them, without knowing the implications for subsequent processing operations and the resulting products. The purpose of this comparison is primarily to evaluate and show the impact of these various processing techniques on the results and the composite image.

For both methods (polynomial and photogrammetric), the processing steps are more or less similar, except for the viewing parameters and the altimetry (ground control points and DEM) involved in the photogrammetric method:

- acquisition of parameters of the viewing geometry (for the photogrammetric method only);
- acquisition of ground control points (GCPs): image coordinates and ground coordinates X, Y, (Z);
- calculation of parameters of the polynomial or photogrammetric model;
- cubic-convolution resampling (with DEM) to create the ortho-images and mosaics, with the same pixel size; and
- registration of the vector file to check the results.

Since the polynomial methods, with their formulation, are well known and documented in [10], only few characteristics are given. The polynomial function of the 1st degree allows the correction of translation, rotation, scaling in both axes and obliquity. Polynomial functions of a higher degree (mainly 2nd and 3rd) enable us to correct larger distortions. However, they are generally limited (small image, flat relief and so on), as they do not reflect the causes of distortions during formation of the image. Moreover, one of the assumptions of these functions is that the ground is flat (with no curvature of the Earth), and without relief.

The photogrammetric model, with its formulation, has been described in detail for different images [8]. This parametric model represents the physical law of transformation from ground space to image space. The development of the final equations is based on principles related to photogrammetry (collinearity condition), orbitography (flight path represented by an osculatory ellipse), geodesy (use of a reference ellipsoid) and cartography (conformity of the projection). It allows integration and combination of the different distortions during image formation, as follows:

- distortions related to the platform (position, velocity, orientation);
- distortions related to the sensor (orientation angles, line integration time, instantaneous field of view);
- distortions related to the Earth (geoid-ellipsoid), including relief; and
- distortions related to the map projection (ellipsoid-map plane).

The main characteristics and comparison of the two processing methods (polynomial and photogrammetric) are summarised in Table 1.

1.2. RADIOMETRIC PROCESSING

There are a number of methods for merging spectral information from different images [11]:

- red-green-blue coding (RGB);
- principal component analysis (PCA);
- intensity-hue-saturation coding (IHS); and
- high-pass filter (HPF).

TABLE 1. Comparison of characteristics for polynomial and photogrammetric methods.

Polynomial Method	Photogrammetric Method
Does not respect the viewing geometry	Respects the viewing geometry
Not related to distortions	Reflects the distortions
Does not introduce attitude data	Uses ephemeris and attitude data
Does not use DEM	Uses DEM or near elevation
Corrects image locally at the GCPs	Corrects the image globally
Does not filter blunders	Filters blunders with the knowledge of the geometry
Individual adjustments of one image	Simultaneous adjustment of more than one image
Image-to-image correction	Image-to-ground correction
Needs many (>20) GCPs	Need few (3-8) GCPs
Sensitive to GCPs distribution	Not sensitive to GCPs distribution
Problem of choice for tie points	GCPs choice as a function of each image

RGB coding is used directly with three images, assigning each ortho-image to a colour: such as SPOT-P in red, SAR-west in green and SAR-east in blue.

The PC method is a statistical method, which transforms by linear combination a data set of variables correlated among themselves into new decorrelated variables. This method generates new orthogonal axes in radiometric space called principal components. The sum of the variance remains unchanged and each consecutive PC has a decreasing level of variance. Depending on the number of images available, the first three PCs are used or one of the PCs can be replaced by another image. In the case of three images, the three resulting PCs of the PCA are used.

IHS coding can be used in two ways:

- the images are used directly to modulate the RGB display of the IHS coding; some authors use the image with the higher spatial resolution, or the SAR, for intensity [12], while others advise modulating saturation rather than intensity [13]; and
- the IHS parameters are calculated on the basis of three images or spectral bands, then one of the parameters is replaced by a fourth image (of higher resolution, or a SAR) and the RGB reverse transformation is performed to merge the images.

Since only three ortho-images are available in our experiment, only the first method of IHS coding is used in the comparisons of the various radiometric merges.

In the HPF method, we use a high-pass filter to process the image with the highest spatial resolution and then combine it, pixel by pixel, with the image having the lowest spatial resolution but the highest spectral resolution. Thus this method combines the spatial information from the image of higher spatial resolution with the spectral information from the image of higher spectral resolution. It applies mainly to combining a panchromatic SPOT-P or SAR image with multiband Landsat-TM or SPOT-XS image. This tool does not apply in our experiment because the images (SPOT-P and airborne SAR) have approximately the same spatial resolution.

2.3. ANALYSIS OF THE RESULTS

2.3.1. *Geometric Processing*

Analysis of the geometric processing results is done in two stages:

- quantitative analysis involving the residuals on the GCPs, the errors on the independent check points (ICPs) and comparison with the vector file;
- qualitative analysis involving a comparison of the cartographic elements (roads, rivers, forest, cutovers and so on) on the two ortho-images.

Table 2, based on 15 GCPs, gives the root-mean-square (RMS) and maximum residuals (in metres) of the calculation of geometric correction models for the photogrammetric method and the polynomial methods (2nd and 3rd orders). Although in the photogrammetric method only four ground control points for SPOT-P and seven for SAR are necessary and the photogrammetric model is not sensitive to the number of GCPs [8], 15 GCPs were used for consistency in the comparison of results between the two techniques.

TABLE 2. Root mean square and maximum residuals (metres) on 15 GCPs for the monoscopic processing.

METHOD	IMAGE Residuals	SPOT-P Rx	Ry	SAR1-EST Rx	Ry	SAR2-EST Rx	Ry	SAR1-EAST Rx	Ry	SAR2-EAST Rx	Ry	ALL Rx	Ry
Photogram-metric	RMSR	2.7	3.1	1.0	2.5	6.3	4.9	6.1	6.1	2.9	4.3	3.5	3.6
	Rmax	-5.0	7.6	- 1.7	4.4	-10.2	-9.9	10.0	10.6	6.4	10.6	7.8	-7.3
Polynomial 2nd order	RMSR	23.1	3.4	4.9	3.7	13.8	4.4	14.0	6.3	13.2	5.1	-	-
	Rmax	-50.3	6.0	-10.8	7.4	-20.8	11.6	-29.6	-9.3	-26.3	-9.7	-	-
Polynomial 3rd order	RMSR	18.7	1.3	4.6	2.6	9.5	3.6	10.1	6.1	9.9	4.5	-	-
	Rmax	-40.6	-2.4	- 9.4	6.2	-20.2	7.4	-27.3	-7.8	-23.1	-8.5	-	-

As it can be noticed in Table 2, the residuals are better for the photogrammetric method than for the polynomial methods. In the X direction, the deviation is more visible because of the elevation distortions, which are modelled in the photogrammetric method. In addition, this method allows simultaneous adjustment of all images by using common points on two or more images as tie points (coplanarity condition). This simultaneous adjustment provides better relative accuracy between the images.

In the photogrammetric method, the residuals are a good indicator of the final accuracy [8], since the correction model is one that corrects the image globally. This is not the case with the polynomial methods, which correct locally at the ground control points. It implies that distortions between the GCPs are not rigorously modelled, and consequently not entirely eliminated.

The fact that the residuals of the 3rd-order polynomial method are better than those of the 2nd order does not imply better accuracy. In the 3rd order, in fact, as there are eight additional unknowns and the same number of GCPs, the degree of freedom in the least squares adjustment is smaller, and thus reduces the adjustment residuals. Since we know the value of the 3rd-order unknowns calculated for each image, we can determine their effect on the ground or their contribution in the correction:

- for SPOT-P, we have: $3.7 \ 10^{-13} \times 6 \ 000^3 < 0.1$ m;
- for SAR, we have: $2.5 \ 10^{-12} \times 4 \ 096^2 \times 10 \ 000 < 0.5$ m;
$4.5 \ 10^{-15} \times 10 \ 000^3 < 0.01$ m.

These 3rd-order parameters are negligible and have no effect in the correction. Despite the results of the residuals, the 3rd-order polynomial does not allow better correction of the images. Moreover, the errors calculated on about twenty ICPs plotted on the ortho-images, are greater (10-20 m) with the 3rd-order polynomial method than with that of the 2nd order. For these reasons, the analysis of the results and the comparison of the ortho-images and their merging will not take the 3rd order into consideration.

Table 3 gives the root-mean-square errors, maximums and bias calculated on about fifty ICPs for the photogrammetric and 2nd-order polynomial methods. These ICPs, plotted on the ortho-images, are different from the 15 GCPs used in calculating the geometric correction models. These errors therefore reflect the final accuracy of the products.

TABLE 3. Root mean square, maximum and bias errors (metres) on 50 check points for the monoscopic processing.

METHOD	IMAGE Errors	SPOT-P Ex	Ey	SAR1-WEST Ex	Ey	SAR2-WEST Ex	Ey	SAR1-EAST Ex	Ey	SAR2-EAST Ex	Ey
Photogram-metric	RMSE	3.8	3.4	5.0	4.3	10.9	6.6	9.1	9.0	7.5	7.6
	Emax	-8.7	-9.9	-11.7	8.5	-24.4	-20.2	23.7	-22.8	-17.6	15.1
	Bias	1.4	-0.1	0.2	0.0	0.3	0.3	4.1	-1.8	0.3	-1.2
Polynomial 2nd order	RMQE	30.0	16.3	13.3	10.3	21.8	14.6	15.7	10.0	21.4	9.1
	Emax	-68.1	31.8	-35.7	-25.2	-46.0	35.1	46.9	27.5	-61.8	-17.5
	Bias	-11.8	11.5	-1.8	-1.4	3.8	-2.9	3.9	-0.2	3.4	-3.3

In any case, the photogrammetric method gives better results than the polynomial method. Note that, for SPOT-P, the differences between the two methods are significantly greater, since modelling of the satellite orbit with the ephemeris is much more accurate than modelling of the aircraft flight with only approximate values for altitude, direction and speed. As in Table 2, the differences are still greater in the X direction, primarily because of the altimetry effects, which are not corrected in the polynomial method.

The SAR-west and SAR-east mosaics and integration of the three ortho-images will therefore be achieved with an absolute error of:

- 10-15 m in the X and Y directions for the photogrammetric method; and
- 30-40 m in the X and Y directions for the polynomial method.

The qualitative evaluation of these geometric processing techniques is performed on the ortho-images and on the colour composite, which has been generated with the IHS coding. Figure 2 is a comparison of two composite subortho-images (4 by 3 km; pixel of 5 m) by the photogrammetric method (top) and by the polynomial method (bottom) to which the road vector file (accuracy of 3-5 m) has been registered. The radiometric processing performed are the same for both images.

The top image is much more homogeneous in its colours, surfaces and variations. As there is greater contrast between elements, their boundaries are clear and well defined. In the bottom image, the colour variations are greater, giving an impression of texture, and the image seems more blurred. As there is less contrast between the elements, they appear less well defined. Using the digital vector file from 1:50,000 scale topographic map, the analysis of some cartographic elements showed, in the bottom image (letters a, b and so on refer to parts of the image identified in Figure 2), that:

(a) the linear elements (roads and rivers) are either doubled or disappear (bridge, roads), due to co-registration error;
(b) the lack of sharpness in this area prevents from distinguishing the road from the forest and areas of bared soil;
(c) on surface elements, artefacts are created; there is an inversion between forest (green) and cutovers (burgundy);
(d) the texture and colour variations do not correspond to the real mapping information.

These examples, with other similar ones, clearly show that the geometric registration errors have generated radiometric merging errors, artefacts and erroneous information in the composite image. These errors do not correspond to any true information related to the ground. The road vector file, registered to these subimages, allows us to check the geometric accuracy: the visual analysis confirms the earlier statistical results for the polynomial method (30-50 m), but shows an improvement for the photogrammetric method (10 m), with maximum errors of 20 metres. These values correspond to the absolute error of registration. Validations on other areas of the image show the consistency of the results.

To confirm the quality of a rigorous geometric processing applied to various remote-sensing images, Figure 3 displays a mosaic of the eight ortho-images with the road network overlaid. The image is 39 by 29 km large with a common 10-m pixel size. From west to east, or north to south, there are the airborne SAR (C-HH), airborne CCD-MEIS sensor, SPOT-P, ERS-1-SAR (C-VV), SEASAT-SAR (L-HH), SPOT-XS (Band 2), Landsat-TM (Band 3) and MOS-MESSR (Band 2). The mosaic becomes fuzzy when viewed diagonally from the 4-m (airborne data) to the 50-m (MOS-MESSR) pixel size, resampled at 10 metres.

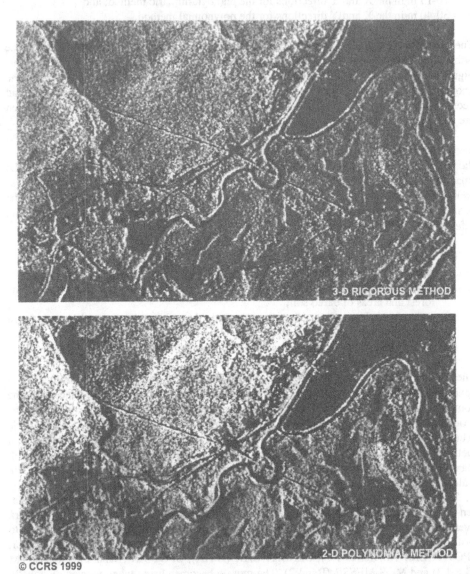

Figure 2. Comparison of two composite subortho-images (4 by 3 km; pixel of 5 m) by the photogrammetric method (top) and by the polynomial method (bottom), to which the road vector file (accuracy of 3-5 m) has been registered. The radiometric processing performed are the same for both images. Please see appendix for image in colour

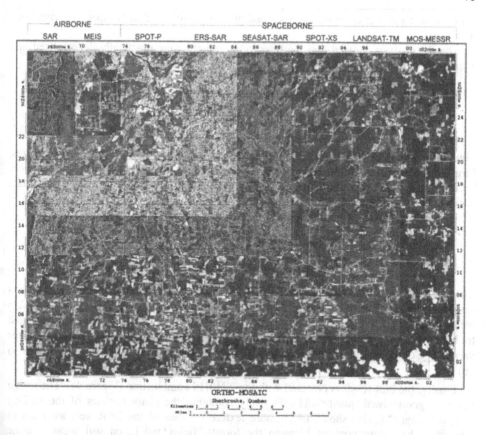

Figure 3. Mosaic of eight ortho-images (39 by 29 km10-m pixel size) with the road network overlaid. From west to east, or north to south, there are the airborne SAR (C-HH), airborne CCD-MEIS sensor, SPOT-P, ERS-1-SAR (C-VV), SEASAT-SAR (L-HH), SPOT-XS (Band 2), Landsat-TM (Band 3) and MOS-MESSR (Band 2). The mosaic becomes fuzzy when viewed diagonally from the 4-m (airborne data) to the 50-m (MOS-MESSR) pixel size, resampled at 10 metres.

2.3.2. *Radiometric Processing*

As the analysis of geometric processing steps has shown that the polynomial methods affect the geometry and radiometry of the composite image, the radiometric processing steps described in 2.2 are only performed on the ortho-images geocoded by the photogrammetric method. Furthermore, only the best composite image is presented in this paper.

RGB coding is used directly by assigning SPOT-P to red, SAR-west to green and SAR-east to blue. In this combination, the characteristics of each image (SPOT-P, SAR) are well preserved. The highly visible elements on SPOT-P come out in red, and the elements oriented west and east come out in green and blue respectively. This is especially visible on river banks.

The PC analysis showed that the three ortho-images were practically decorrelated and that:

- the first PC is 99% formed of SPOT-P;
- the second PC is 97% formed of SAR-west; and
- the third PC is 97% formed of SAR-east.

Thus using the three PCs contributes no more than using the three original ortho-images. Moreover, the results are often more difficult to interpret quantitatively and qualitatively because, as the statistical properties have been manipulated, the original integrity of the data has not been preserved [14].

Different IHS coding were tested and the two best one were:

- SPOT-P in I, SAR-west in H and SAR-east in S; and
- SAR-west in I, SPOT-P in H and SAR-east in S.

The first combination somewhat resembles a colour air photo since the visible SPOT-P was assigned to the intensity, which represents the brightness of colour. The highly visible elements on SPOT-P come then out very well in bright colour. As SAR-west was assigned to hue, which represents the dominant colour, it does not help provide much colour variation. Consequently, many characteristics of SAR are not visible (texture, relief and so on).

Finally, the best result is obtained with the 2nd IHS combination (Figure 4). The image has very good visual quality and effectively combines the characteristics of the various original images. It also shows much more texture because of the SAR-west assigned to intensity. The colour contrast between the forests, fields and bared soil areas is quite pronounced. This last combination seemed to be the most logical in our case, since SPOT-P covers the visible spectrum, and the higher-resolution SAR images (4 m versus 10 m), with more texture, better modulate intensity and saturation. It corresponds to tests and results of Jaskolla et al. [12] and Welch and Ehlers [13].

2.4. TOPOGRAPHIC MAPPING

To evaluate the mapping potential, AN image content analysis and visual interpretation of the best composite image using HIS radiometric processing (Figure 4) is performed with regards to the conventional applications of remote sensing: cartography, agriculture, forestry and geology.

Figure 4. Composite ortho-images (10 by 10 km; 5-m pixel spacing) using IHS radiometric coding with SAR-west in I, SPOT-P in H and SAR-east in S. Please see appendix for image in colour.

2.4.1. *Cartography*

Roads can be distinguished easily because of the spatial resolution (5 m) and the contrast with other elements, such as the buildings and built-up areas. Similarly, the roads in new residential developments in forested areas are clearly visible in this image. For rivers, there is little colour variation from the SPOT-P and the moderate contrast only allows us to distinguish the boundaries. Finally, the shadows and their orientation are enhanced by the use of two SARs of opposite viewing directions; moreover, the coding of the SAR-west mosaic in intensity accentuates the texture of the image.

2.4.2. *Agriculture*

The boundaries of fields are clearly visible. These boundaries are enhanced by fences, which are highly visible because of the prominent SAR information in these images. For the same reason, fields containing stumps or undergoing reforestation are identifiable. As the dynamic range is great, it also allows better discrimination between land uses and between bared and cultivated fields.

2.4.3. *Forestry*

The image is very good for distinguishing forest from everything else. However, it is practically impossible to distinguish between deciduous and coniferous trees. This must come from the SPOT-P intensity image, since conifers are darker in SPOT-P images. Texture on the tree canopy related to the size of the crown and not to tree type (deciduous versus coniferous) can de discriminated. Rows of isolated trees are also visible because of their shadow. There is a visual impression of tree height superimposed on the relief, allowing us to interpret the characteristics and disturbances of stands on the basis of forest cover height. Moreover, this impression, combined with the shading, lets us distinguish rows of isolated trees.

2.4.4. *Geology*

When information on the relief is not useful, this image easily allows the distinction of more or less the same geomorphologic elements: the two NE-SW rivers and their characteristics (meanders, embankments, and bars). As soon as the interpretation requires knowledge of the relief, this composite image is much more useful due to the relief perception: stream bank slopes and glacial formations, with drumlins and ridges, which indicate the NE-SW ice advance. Similarly, NE-SW lineaments and folds, identifiable only on these two images, are probably related to the structural trend of the region.

3. Stereoscopic Processing

When no DEM is available and two images from the same sensor (VIR or SAR) are available, the stereoscopic method for feature extraction is based on traditional photogrammetric techniques. Even with two images from different sensors, the brain can generate the perception of depth, combining for example the spectral information from the Landsat-TM image and the spatial information from the SPOT-P image for the stereo plotting. The XY cartographic coordinates of the planimetric features are computed independently of its Z-altimetric coordinate, since the operator always plots in stereoscopy at the vertical of the point [15]. Consequently, the planimetric accuracy of feature positioning is not affected by any error on elevation, conversely to the previous method where any error in the DEM propagates through the geocoding process and the planimetric features.

The processing steps are more or less similar to the monoscopic method, except for the viewing parameters and the altimetry (ground control points and DEM) involved in the photogrammetric method:

- acquisition of parameters of the viewing geometry;
- acquisition of GCPs in stereoscopy: image coordinates and ground coordinates X, Y, Z;
- calculation of parameters of the stereoscopic model; and
- 3D-data extraction on the "virtual" stereo model.

Whatever the data (VIR and SAR), most of the research studies on stereoscopy around the world have focused on DEM generation [16, 17] for the topography, and very few on planimetric features extraction for cartography. Consequently, only results on DEM generation from VIR scanners and SAR sensors are presented.

3.1. VIR SCANNERS

To obtain stereoscopy with images from satellite scanners, three solutions are possible:

- the adjacent-track stereoscopy from two different orbits;
- the across-track stereoscopy from two different orbits; and
- the along-track stereoscopy from the same orbit using fore and aft images.

3.1.1. Adjacent-track

In the case of Landsat (MSS or TM) and Indian IRS-1A satellites, the stereoscopic acquisition is only possible from two adjacent orbits since the satellite only acquires nadir viewing images, and the tracking orbit ensures repeat path consistent within a few kilometres [18]. In fact the B/H ratio with Landsat-MSS is around 0.1, so that relief of about 4 000 m is needed to generate a parallax of five Landsat-MSS pixels (80-m resolution). Due to its quasi-polar orbit, the coverage overlap grows from about 10% at the Equator to about 85% at 80° latitude. From 50° north and south the coverage overlap (45%) enables quasi-operational experiments for elevation extraction and the accuracy of derived DEM is in the order of 50-100 m.

Consequently, the stereoscopic capabilities and applicabilities of "adjacent orbit" satellite data still remain limited because:

- it can be used for large area only in latitude higher than 45° to 50° north and south;
- it generates a small B/H ratio leading to elevation errors of more than 50 m; and
- only medium to high relief areas are suitable for generating enough vertical parallaxes.

3.1.2. Across-track

To obtain good geometry for a better stereo plotting, the intersection angle should be large in order to increase the stereo exaggeration factor, or equivalently the observed parallax, which is used to determine the terrain elevation. B/H ratios of 0.6 to 1.2 are typical values to meet the requirements of topographic mapping [19]. There are only few operational satellites, which have this capability to generate such B/H ratios:

- The SPOT system by steering the sensor (±26°); and
- The IRS-1C/D system by rolling the satellite (±20°).

Since the advent of the SPOT system in 1985, it is the most popular stereo capability and numerous researches around the world were performed [16]. They lead to accuracy in elevation from one to few pixels [20] depending on processing methods (photogrammetric or non-parametric), systems (intensity or feature matching) and tools (automatic, semi-automatic) used.

The new high-resolution IKONOS system, launched in September 1999, should be able to also provide across-track stereo-images, but in addition it has along-track stereo-capability by steering the sensor in any direction (±26°). If the raw data is available to end users it should confirmed the previous results achieved with SPOT data.

3.1.3. *Along-track*

In the last few years, the last solution as applied previously to space frame cameras got renewed popularity. First, the JERS-1's Optical Sensor (OPS) [21] and the German Modular Opto-Electronic Multi-Spectral Stereo Scanner (MOMS) [22] generate stereo-images by the use of forward and nadir linear array optical sensors, named OPS. The 15° forward-looking image and the nadir-looking image (18-m ground resolution) generate a stereo-pair with a B/H ratio of 0.3. The simultaneous along-track stereo-data acquisition gives a strong advantage in terms of radiometric variations versus the multi-date stereo-data acquisition with across-track stereo. This was confirmed by the very high correlation success rate (82.6%) [23]. However, the limited availability of data has restricted the evaluation of DEM generation to few research groups. They obtained accuracy for DEM of about few pixels, which are generally not as good as those obtained with across-track stereo-images.

In the next future, the Advanced Spaceborne Thermal Emission and Reflection Radiometer (ASTER) [24], the Indian IRS-P5, and most of the high-resolution satellites such as Orb-View1 and Quick-Bird and IKONOS will also be a good data source and enable better evaluations. Preliminary evaluation using aerial imagery scanned at 1-m spatial resolution showed their potential to obtain a RMS error in elevation in the range of 1.5 m to 2 m [25]. However, it is not sure that the raw imagery needed for generating DEMs and derivative topographic products will be available to the end users since, at that time, the high-resolution data resellers want to only distribute value-added products (DEM, ortho-images, mosaics).

Table 4 summarises the general results of DEM extraction with different VIR scanners using the three stereoscopic methods. Some variations in the results occur due mainly to the different geometric modelling, image matching, editing, digital or not processing.

TABLE 4. Summary of the results of the elevation extraction with the VIR scanners using the stereoscopic method. The variations in the results for each stereo configuration are due to the different research studies. The values in brackets were obtained from simulated data.

Stereo-Pairs	Resolution	Adjacent-track	Across-track	Along-track
Landsat MSS	80 m	100-300 m		
Landsat TM	30 m	45-70 m		
IRS 1A	72 m	35 m		
SPOT P	10 m		5-15 m	
SPOT/Landsat	10 m/30 m		35-50 m	
IRS 1C/D	6 m		10-30 m	
MOMS-2	13.5 m			5-15 m
MOMS-2P	18 m			10-30 m
JERS OPS	20 m			20-40 m
SPOT/ERS	10 m/30 m		20-30 m	
EOS-ASTER	33 m		(15 m)	(12.5 m)
Ikonos	1 m		(1.5-2 m)	

3.2. SAR SENSORS

Due to the specific geometric and radiometric aspects of SAR images, it may take our brain time to perceive the terrain relief with SAR stereo-images, mainly when both geometric and radiometric disparities are large [26]. However, since depth perception is an active process (brain and eye) and relies on an intimate relationship with object recognition, radar images can be viewed in stereo as easily as VIR satellite images after training. Stereo parallaxes predominate when viewing radar images, but the shade and shadow cues also have a strong and cumulative effect. For example, with a quasi-flat terrain, the shade and shadow cues overcome the stereo effect when viewing pseudoscopically a radar stereopair [27].

To obtain good geometry for stereoplotting, the intersection angle (Figure 5) should be large in order to increase the stereo exaggeration factor, or equivalently the observed parallax, which is used to determine the terrain elevation. Conversely, to have good stereo-viewing, the interpreters (or the image matching software) prefer images as nearly identical as possible, implying a small intersection angle. Consequently, large geometric and radiometric disparities together hinder stereo-viewing and precise stereoplotting. Thus, a compromise has to be reached between a better stereo-viewing (small radiometric differences) and more accurate elevation determination (large parallax).

The common compromise for any type of relief is to use a same-side stereopair, thus fulfilling both conditions above. It was realised with SIR-B [28], SIR-C [29], ERS [30] and ERS [23]. Unfortunately, this does not maximise the full potential of stereo radar for terrain relief extraction. Another potential compromise is to use opposite-side stereopairs over rolling topography [15]. The rolling topography reduces the parallax difference and also the radiometric disparities (no layover and shadow, little foreshortening) making possible simultaneously good stereo-viewing and accurate stereoplotting. A last approach to minimise the geometric disparities is to pre-process the images using a large grid spacing or low accuracy DEM, as it has been applied with success to iterative hierarchical SAR image matching [28].

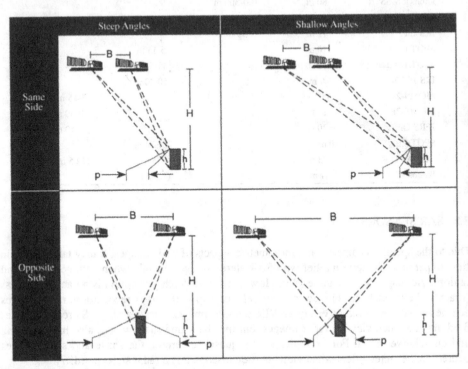

Figure 5. The intersection geometry with the radar parallax (p) due to the terrain elevation (h) for different stereo SAR configurations (same-side versus opposite-side; steep versus shallow look-angles).

Since the last ten years, most of the results on DEM generation with SAR stereo-images have been inconsistent and practical experiments do not clearly support theoretical expectations [31]. For example, larger ray intersection angles and higher spatial resolution do not translate into higher accuracy. In various experiments, accuracy trends even reverse, especially for rough topography. Only in the extreme case of low relief, does accuracy approach the theoretical expectations. The main reason is that the error modelling accounts only for SAR geometric aspects (look and intersection angles, range error) and completely neglects the radiometric aspects (SAR backscatter) of the stereopair and of the terrain.

Since SAR backscatting, and consequently the image radiometry, is much more sensitive to the incidence angle that the VIR reflectance, especially at low incidence angles, the possibility of using theoretical error propagation as a tool for predicting accuracy and selecting appropriate stereo-images for DEM generation is very limited. Therefore, care must be taken in attempting to extrapolate VIR stereo concepts to SAR.

Previously to RADARSAT, Canada's first earth observation satellite launched in November 1995, it was difficult to acquire different stereo configurations to address the above points. RADARSAT with its various operating modes, imagery from a broad range of look directions, beam positions and modes at different resolutions [32] fills this gap. Under the Applications Development and Research Opportunity (ADRO) program sponsored by the Canadian Space Agency, researchers around the world have undertaken studies on the stereoscopic capabilities by varying the geometric parameters (look and intersection angles, resolution, etc.). Most of the results were presented at the final RADARSAT ADRO Symposium "Bringing Radar Application Down to Earth" held in Montreal, Canada in 1998 [17]. There was a general consensus on the achieved DEM extraction accuracy: a little more (12 m) and a little less (20 m) than the image resolution for the fine mode and the standard mode respectively, whatever the method used (digital stereoplotter or image matching). Relative elevation extraction from a fine mode RADARSAT stereopair for the measurement of canopy heights in the tropical forest of Brazil was also addressed [27].

However, there were no significant correlations between the DEM accuracy and the intersection angle in the various ADRO experiment results. This confirmed the same contradiction found with SIR-B [29]. In fact, most of the experiments showed that the principal parameter that has a significant impact on the accuracy of the DEM is the type of the relief and its slope. The greater the difference between two look-angles (large intersection angle), the more the quality of the stereoscopic fusion deteriorated. This cancels out the advantage obtained from the stronger stereo geometry, and is more pronounced with high-relief terrain. On the other hand, although a higher resolution (fine mode) produced a better quality image, it does not change the stereo acuity for a given configuration (e.g. intersection angle), and it does not improve significantly the DEM accuracy. Furthermore, although the speckle creates some confusion in stereoplotting, it does not degrade the DEM accuracy because the matching methods or the human stereo-viewing "behave like a filter". Preprocessing the images with an adaptive speckle filtering does not improve the DEM accuracy with a multi-scale matching method [30]; it can slightly reduce the image contrast and smoothes the relief, especially the low one [26].

Since the type of relief is an important parameter influencing the DEM accuracy, it is strongly recommended that the DEM accuracy be estimated for different relief types. Furthermore, in the choice of a stereoscopic pair for DEM generation, both the geometric and radiometric characteristics must be jointly evaluated taking into account the SAR and surface interaction (surface geometry, vegetation, soil properties, geographic conditions, etc.). The advantages of one characteristic must be weighted against the deficits of the other. Table 5 summarises the general results of DEM extraction with SAR scanners using the stereoscopic method.

TABLE 5. Summary of the results of the elevation extraction with the SAR sensors using the stereoscopic method. The variations in the results for each stereo configuration are due to the different research studies.

Satellite	SAR Band-Polarisation	Resolution (m)	Relief	Accuracy (m)	
				Same-Side	Opposite-Side
SIR A	L-HH	25	High	100	
SIR B	L-HH	40	Medium	25	
			High	60	36
ERS 1/2	C-VV	24	Medium	20	20
			High	45	
JERS	L-VV	18	High	75	
Almaz	S-HH	15	High	30-50	
		P[a] 7-9	Low	8-10	20
RADARSAT	C-HH	S[a] 20-29	Medium	15-20	40
		W[a] 20-40	High	25-30	

4. Conclusion

Two conventional methods (monoscopic and stereoscopic) to process remote sensing images for extracting 2D or 3D information have been presented.

The monoscopic method requires rigorous geometric and radiometric processings. The superiority of the rigorous geometric processing is mainly due to the fact that the mathematical model corresponds to the physical reality of the viewing geometry and takes into account the distortions caused by relief. This superiority will also increase with mountainous terrain. This rigorous geometric processing will facilitated subsequent processing operations, while the polynomial geometric processing will require more complicated processing operations to remove the artefacts and false information. Furthermore, because the latter do not correspond to any physical reality and depend on viewing conditions (images, ground and so on), the subsequent processing techniques are dependent on the specific viewing conditions, and will not apply with another set of images under different viewing conditions. It thus limits the use and future applications of such image processing techniques.

Consequently, the monoscopic processing of multisource data requires rigorous geometric correction to obtain a subpixel accuracy, as well as appropriate radiometric processing, which take into account the nature and characteristics of the data. It then ensures that the composite image preserves the best of the information from each image and maintains data integrity.

On the other hand, the stereo capability of different satellites with different methods has been addressed: adjacent-orbit stereo with Landsat and IRS-1A, across-track stereo with

SPOT and IRS-1C, along-track stereo with JERS and MOMS, same and opposite-stereo with ERS and RADARSAT.

Since any sensor, system or method has its own advantages and disadvantages, future solution for operational DEM generation should use the complementarity between the different sensors, systems, methods and processing. Furthermore, it has been proven in most of the previous experiments that the user has to make judgements and decisions at different stages of the processing, regardless of the level of automatic processing to obtain the final DEM product. Non-exhaustive examples of complementarity are listed below:

- to use mixed-sensor (VIR and SAR) stereoscopic images in order to obtain the second image of the stereo-pair in cloud-cover area;
- to combine VIR and SAR stereoscopic images where the radiometric content of the VIR image is combined with the SAR high sensitivity to the terrain relief and its "all-weather" capability;
- to use the visual matching to seed points to the automatic matching or to post-process and edit raw DEMs (occlusion, shadow or mismatch areas);
- to use stereo measurements of objects edges and other geomorphological features (thalweg and crest lines, break lines, lake boundary and elevation) to increase the consistency of the DEM;
- to combine the "know-how" of the users with the computer capability.

5. Acknowledgements

The author would like to thank NATO and CCRS in their support. Ms. Liyuan Wu and Mr. René Chenier of Consultants TGIS inc., who performed image processing, are also acknowledged.

6. References

1. Guertin, F. and Shaw, E. (1981) Definition and potential of geocoded satellite imagery products, *Proceedings of the 7th Canadian Symposium on Remote Sensing*, held in Winnipeg, Canada, September 8-11, Manitoba Remote Sensing Centre, Canada, 384-394

2. Clark, J. (1980) Training site statistics from Landsat and Seasat satellite imagery registered to a common map base, *Proceedings of the ASPRS Semi-Annual Convention*, Niagara Falls, U.S.A., American Society of Photogrammetry, RS.1.F.1-RS.1.F.9

3. Murphrey, S.W. (1978) SAR-Landsat image registration study, Final report, IBM Corp., Gaithersburg, MD, USA.

4. Anuta, P.E., Freeman, D.M., Shelly, B.M. and Smith, C.R. (1978) SAR-Landsat image registration systems study, LARS Contract Report 082478, Purdue University, Ind., U.S.A.

5. Daily, M., Farr, T., Elachi, C. and Schuber, G. (1979) Geologic interpretation from composite radar and Landsat imagery, *Photogrammetric Engineering and Remote Sensing*, 45, 1109-1116.

6. Guindon, B., Harris, J.W.E., Teillet, P.M., Goodenough, D.G. and Meunier, J.F. (1980) Integration of MSS and SAR data for forested regions in mountainous terrain, *Proceedings of the 14th International Symposium of Remote Sensing Engineering* held in San Jose, Costa Rica, ERIM, USA, 79-84.

7. Zobrist, A.L., Blackwell, R.J. and Stromberg, W.D. (1979) Integration of Landsat, Seasat and other geodata sources, *Proceedings of the 13th Annual Symposium on Remote Sensing of the Environment*, ERIM, Ann Arbor, USA, 271-279.

8. Toutin, Th. (1995) Multi-source Data Fusion with an Integrated and Unified Geometric Modelling, *EARSeL Journal Advances in Remote Sensing*, **4**, 118-129.

9. Livingstone, C.E., Gray, A.L., Hawkins, R.K., Olsen, R.B., Halbertsma, J.G. and Deane, R.A. (1987) CCRS C-band airborne radar: system description and test results, *Proceedings of the 11th Canadian Symposium on Remote Sensing*, Waterloo, Canada, 22-25 June, University of Waterloo, Canada, 379-395.

10. Colwell, R.N. (1983) *Manual of Remote Sensing*, 2nd edition, Vol. 1, Sheridan Press, Falls Church, Virginia, U.S.A.

11. Chavez, P.S. Jr., Sides, S.C., and Anderson, J.A. (1991) Comparison of three different methods to merge multiresolution and multispectral data: Landsat-TM and SPOT panchromatic, *Photogrammetric Engineering and Remote Sensing*, **57**, 295-303.

12. Jaskolla, F., Rast, M., and Bodechtel, J. (1985) The use of SAR system for geological applications, *Proceedings of the Workshop on Thematic Applications of SAR Data*, Frascati, Italy, SP-257, ESA, Paris, 41-50.

13. Welch, R. and Ehlers, M. (1988) Cartographic feature extraction with integrated SIR-B and Landsat-TM images, *International Journal of Remote Sensing*, **9**, 873-889.

14. Harris, J.R., Murray, R. and Hirose, T. (1990) IHS transform for the integration of radar imagery with other remotely sensed data, *Photogrammetric Engineering of Remote Sensing*, **56**, 1631-1641.

15. Toutin, Th. (1996) Opposite-side ERS-1 SAR stereo mapping over rolling topography, *IEEE Transactions on Geoscience and Remote Sensing* **34**, 543-549.

16. Centre National d'Études Spatiales (CNES) (1987) *SPOT-1 : Utilisation des images, bilan, résultats*, *Proceedings of the SPOT-1 Symposium*, Paris, France, CNES, Toulouse, France.

17. Canadian Space Agency (CSA), 1998, "Bringing Radar Application Down to Earth", *Proceedings of the RADARSAT ADRO Symposium*, Montreal, Canada, October 13-15, CD-ROM.

18. Simard, R., 1983, "Digital stereo-enhancement of Landsat-MSS data", *Proceedings of the Seventeenth International Symposium on Remote Sensing of Environment*, ERIM, Ann Arbor, MI, USA, May 9-13, 1275-1281.

19. Light, D.L., Brown D., Colvocoresses A., Doyle F., Davies M., Ellasal A., Junkins J., Manent J., McKenney A., Undrejka R. and Wood G. (1980) Satellite photogrammetry, Chapter XVII in *Manual of Photogrammetry*, ASPRS, Bethesda, USA, pp. 883-977.

20. Toutin, Th. (1995) Generating DEM from stereo images with a photogrammetric approach: Examples with VIR and SAR data, *EARSeL Journal Advances in Remote Sensing*, **4**, 110-117.

21. Maruyama, H., Kojiroi R., Ohtsuka T., Shimoyama Y., Hara S. and Masaharu H. (1994) Three dimensional measurement by JERS-1, OPS stereo data, *International Archives for Photogrammetry and Remote Sensing*, Athens, Ga, USA, **30**, 210-215.

22. Ackerman, F., Fritsch D., Hahn M., Schneider F. and Tsingas V. (1995) Automatic generation of digital terrain models with MOMS-02/D2 data, *Proceedings of the MOMS-02 Symposium*, Köln, Germany, July 5-7, EARSeL, Paris, France, 79-86.

23. Raggam, J. and Almer A. (1996) Assessment of the potential of JERS-1 for relief mapping Using optical and SAR data, *International Archives of Photogrammetry and Remote Sensing*, Vienna, Austria, July 9-18, Austrian Society for Surveying and Geoinformation, Vienna, Austria, **31**, B4, 671-676.

24. Tokunaga, M., Hara S., Miyazaki Y. and Kaku M. (1996) Overview of DEM product generated by using ASTER data, *International Archives of Photogrammetry and Remote Sensing*, Vienna, Austria, July 9-18, Austrian Society for Surveying and Geoinformation, Vienna, Austria, **31**, B4, 874-878.

25. Ridley, H.M., Atkinson P.M., Aplin P., Muller J.-P. and Dowman I. (1997) Evaluation of the potential of the forthcoming US high-resolution satellite sensor imagery at the ordnance survey, *Photogrammetric Engineering and Remote Sensing*, **63**, 997-1005.

26. Toutin, Th. (1999) Error tracking of radargrammetric DEM from RADARSAT images. *IEEE Transactions on Geoscience and Remote Sensing,* **37**, 2227-2238.

27. Toutin, Th. and Amaral S. (2000) Stereo RADARSAT data for canopy height in Brazilian forest. *Canadian Journal for Remote Sensing* **26,** (in press).

28. Simard, R., Plourde F. and Toutin Th. (1986) Digital elevation modelling with stereo SIR-B image data. *International Archives of Photogrammetry and Remote Sensing,* **26**, 161-166.

29. Leberl, F., Domik G., Raggam J., Cimino J., and Kobrick M. (1986) Multiple incidence angle SIR-B experiment over Argentina: stereo-radargrammetric analysis. *IEEE Transactions on Geoscience and Remote Sensing* **24**, 482-491.

30. Dowman, I.J., Twu Z.-G., Chen P.H. (1997) DEM generation from stereoscopic SAR data. Proceedings *ISPRS Joint Workshop on Sensors and Mapping from Space,* Hannover, Germany, September 29-October 2, 113-122.

31. Leberl, F. (1990) *Radargrammetric image processing.* Artech House, Norwood, USA.

32. Parashar, S., Langham E., McNally J., Ahmed S. (1993) RADARSAT mission requirements and concepts, *Canadian Journal of Remote Sensing* **18**, 280-288.

LAND COVER – LAND USE MAPPING WITHIN THE EUROPEAN CORINE PROGRAMME

G. BÜTTNER
FÖMI Remote Sensing Centre, Bosnyák tér 5. Budapest, H-1149 Hungary
C. STEENMANS
European Environment Agency, Kongens Nytorv 6, Copenhagen
DK-1050, Denmark
M. BOSSARD
IGN FI, 39 ter, rue Gay Lussac, Paris, F-75005 France
J. FERANEC
IG SAS, Stefánikova 49, 81473 Bratislava, Slovak Republic
J. KOLÁR
GISAT, Charkovska 7, 10100 Praha 10, Czech Republic

Abstract

The aim of the CORINE Land Cover Mapping is to provide information on the state and changing biophysical coverage of the Earth's surface. The European Union's CORINE (Co-ordination of Information on the Environment) Land Cover Project was initiated in the EU countries in the 80's to provide quantitative, consistent and comparable information on land cover, at a scale of 1:100 000. After the political changes in Central and Eastern Europe, the project has been extended to the East within the frame of the Phare Programme. Today the CORINE Land Cover database covers 31 countries.

CORINE Land Cover is mapped by interpretation of satellite images, and the results are stored as databases in Geographic Information Systems. These databases represent a basic tool for studies on the environment, impact assessment and regional planning on national as well as on European level.

The paper gives a short historical overview about the evolution of the project and introduces its recent institutional background. It is followed by a brief technical overview over the basic methodology and discussion on the two most exciting extensions of the project: land cover change detection and mapping at larger scale. Finally some applications of CORINE Land Cover data are summarised.

1. Evolution of the Project

1.1. CORINE LAND COVER IN THE EUROPEAN UNION

The idea to produce a uniform pan-European land cover database dates back to the early 80's. It has been recognised that land cover is a basic information for the management of the environment and natural resources. Land cover mapping has become an integral part of

89

M.F. Buchroithner (ed.),
Remote Sensing for Environmental Data in Albania: A Stragegy for Integrated Management, 89–100.
© 2000 Kluwer Academic Publishers. Printed in the Netherlands.

the CORINE (Co-ordination of Information on the Environment) Programme, started in 1985 by the European Commission Directorate General XI (EC DGXI) with the main aim to compile consistent and compatible information on the environment for countries of the European Union.

Information provided by earth observation satellites has become a basic data support to produce land cover inventories. Following a feasibility study, basic methodological questions had been clarified (nomenclature, scale, guidelines for visual photointerpretation, etc). As a pilot project, Portugal was the first country that had been mapped between 1986 and 1990. After that the CORINE Land Cover methodology had been finalised and a Technical Guide was produced [8]. In 1994 the European Environment Agency has started its operation in Copenhagen (Denmark), taking over the maintenance and use of the CORINE Land Cover database as well.

The CORINE Land Cover Project has been implemented in most of the EU countries as well as in the Phare partner countries in Central and Eastern Europe, Morocco and Tunisia. In each country local team(s) had implemented the project along with the supervision of the CORINE Land Cover Technical Unit (LCTU). North European countries (Sweden, Finland) and Great Britain have developed specific GIS-based procedures capable to derive CORINE Land Cover classes, starting from automatic classification of satellite imagery [10, 17].

1.2. CORINE LAND COVER IN CENTRAL AND EASTERN EUROPE

Following political changes in Central and Eastern Europe, the CORINE Land Cover had started in 1993 in six countries (Bulgaria, Czech Republic, Hungary, Poland, Slovak Republic and Romania), in the frame of the Phare Regional Environment Programme. They had finished their project in 1996 [15]. Owing to the good project co-ordination provided by LCTU, and enthusiasm of national teams, the quality of these databases is very high. The CORINE Land Cover inventory of the above six countries has been published on a CD-ROM [4].

As continuation of the Phare Programme, five other countries - Albania, Estonia, Latvia, Lithuania and Slovenia - have completed their databases between 1996 and 1999. Two republics of the former Yugoslavia, Bosnia-Herzegovina and Macedonia, will complete the project in early 2000. Table 1 summarises countries and organisations involved in the CORINE Land Cover Project.

1.3. TOPIC CENTRE AND TOPIC LINK ON LAND COVER

The European Environment Agency (EEA), established by the Commission, handles the European Environment Information and Observation Network (EIONET). Institutions or organisations have been contracted as European Topic Centres to execute particular tasks identified in the Agency's multi-annual work programme, one of which is the European Topic Centre on Land Cover (ETC/LC). The main objective of the ETC/LC is to produce, provide and manage land cover information for environmental policy development and implementation in Europe[16].

During the first 2 years of operation, the ETC/LC concentrated on the completion of the CORINE Land Cover database for the EU member states, on merging with the

Central and Eastern European CORINE Land Cover inventory and gearing-up the
ETC/LC network activities.

TABLE 1. Status of CORINE Land Cover Mapping [5, 13]

Country	area (km²)	contractor	status	no. of sheets
Albania	29 000	IGN FI	completed in 1999	36
Austria	84 000	Umweltbundesamt	completed in 1996	35
Belgium	31 000	IGN/NGI	completed in 1995	22
Bosnia-Herzegovina	51 000	GZ BiH	completed in 2000	38
Bulgaria	111 000	Ministry of Environment	completed in 1996	105
Czech Republic	79 000	GISAT	completed in 1996	94
Denmark	44 000	AFA	completed in 1994	37
Estonia	45 000	EEIC	completed in 1998	37
Finland	337 000	FEI	completed in 1999	342
France	550 000	IGN FI	completed in 1996	311
Germany	357 000	Stat. Bundesamt	completed in 1996	266
Greece	132 000	HEMCO	completed in 1995	134
Hungary	93 000	FÖMI	completed in 1996	84
Ireland (incl. N. Ireland)	84 000	OSI, OSNI	completed in 1993	15
Italy	302 000	ITA Consortio	completed in 1996	278
Latvia	64 000	LEDC	completed in 1998	44
Lithuania	65 000	HNIT-Baltic	completed in 1998	42
Luxemburg	2 600	Walphot, G²ERE	completed in 1990	1
F.Y.R.O. Macedonia	25 000	Ministry of Urban Planning, Construction and Environment	completed in 2000	22
Netherlands	42 000	SC-DLO	completed in 1992	8
Poland	313 000	IGIK	completed in 1996	297
Portugal	92 000	CNIG	completed in 1990	53
Romania	238 000	IGR	completed in 1996	196
Slovak Republic	49 000	IG SAS	completed in 1996	55
Slovenia	20 000	GZS	completed in 1998	20
Spain	505 000	IGN	completed in 1991	296
Sweden	450 000	SSC	started in 1998	225
United Kingdom	240 000	ITE	started in 1998	204
Total: 28	4 434 600			3 297

Since July 1997, the EEA activities on Land Cover are now extended with support
from the DGIA Phare Programme towards Central and Eastern European countries by
the creation of the Phare Topic Link on Land Cover (PTL/LC). This body, consisting of
four institutions from Phare countries has been contracted to provide and develop land
cover information for and about Phare countries (http://www.gisat.cz/ptl). With the set-
up of the PTL/LC, this European network for land cover information exchange is now
extended to 31 countries. This Extended ETC/LC works on one work programme for

monitoring, data and information gathering, analysis and reporting of land cover and land use related issues at a European scale. The Extended ETC/LC concentrates on three main tasks:

- the support and development of policy relevant European environmental applications and indicators for land cover;
- the maintenance and update of the European CORINE Land Cover database and the linking with other European and international networks;
- land cover and land use change analysis at European level.

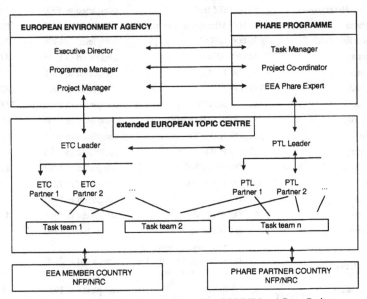

Figure 1. Institutional connections of the CORINE Land Cover Project.

As an integral part of those tasks, the ETC/LC is providing support to the Commission and other ETCs in the integration of land cover data with other environmental data for European applications, mainly in the domains of nature conservation, land planning and water management. The ETC/LC concentrated during 1998 on following EC policy relevant issues:

- support to DGVII (Transport) and DGXI (Environment and Nuclear Safety) on the strategic environmental assessment of the trans-European transport network for revision of the guidelines in 1999;
- support to DGXVI (Regional Development) on the development of the European Spatial Development perspective in 1998;
- support to the ETC on Nature Conservation for the preparation of NATLAN, which is a new EEA key product for 1998 for presentation and dissemination of nature and land cover related information;

- support to the ETC on Inland Waters on the development of the DGXI proposed water resources framework by the use and integration of land cover data in a Geographic Information System (GIS) environment.
- the specific role of PTL/LC in fulfilling the tasks of the Extended ETC/LC is:
- to produce a retrospective CORINE Land Cover database of the 70's in Central and Eastern Europe using archived Landsat MSS images;
- to analyse and report major land cover and land use changes, which took place on the 2nd level of the nomenclature in a period of twenty years [14];
- to assist ETC/LC in its issues of support to the Commission if appropriate, including co-operation with other PTLs.

2. Technology of CORINE Land Cover Mapping

2.1. ELEMENTS OF THE METHODOLOGY

The basic aim of the CORINE Land Cover Project is to provide an inventory of the Earth surface features for the management of the environment. The approach of computer-assisted visual interpretation of satellite images has been chosen as mapping methodology. The choice of the scale (1:100 000), the minimum area to be mapped (25 hectares) and the minimum width of linear elements (100 metres) represent a trade-off between cost and detail of land cover information.

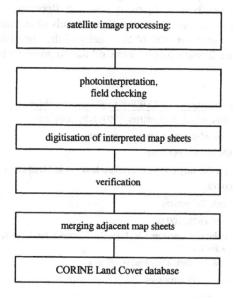

Figure 2. Flowchart of CORINE Land Cover Mapping

Raw satellite images first have to be pre-processed and enhanced to yield a geometrically correct document (satellite image map) in the desired map projection. Digital film-writers are used as output device to produce an optimal hardcopy for visual photointerpretation. Recently large-sized, high-resolution computer printers are also becoming a selectable alternative for output production. In the course of interpretation, the photointerpreter draws polygons on transparent overlay, fixed on the top of a satellite image hardcopy. The photointerpreter should simultaneously consult all available ancillary data (topographic maps, aerial photographs, vegetation maps, land cover and land use statistics, etc.). The polygons are coded according to one of the items (a three-digit code) of the nomenclature. Although display and evaluation of satellite imagery on the screen of an image processing system was highly recommended to compensate for drawbacks of using hardcopy images, it depended on availability of appropriate facilities. In recent times, however, with the rapid spread of lower cost hardware and software systems, computer-assisted photointerpretation (CAPI) technology is more and more applied and provides better thematic and positional accuracy.

Landsat Thematic Mapper (TM) data have been the most widely used satellite imagery in the course of the project, because it provides good area coverage, sufficient thematic and geometric details for a reasonable cost. Recently, newer high-resolution satellites with a middle infrared channel (IRS-1C/D, SPOT-4) certainly increase the possibilities.

Field checking is an integral part of the project methodology either for resolving ambiguities or for general examination of the photointerpretation results. To ensure consistency and comparability the technical guidebook [8] includes instructions and examples for each step of the methodology.

The result of photointerpretation is then digitised and stored in a topologically structured format. The final product is a digital land cover database in a GIS. At the beginning of the project mostly table digitising was used. Recently, most teams use raster scanning and a subsequent vectorising, with different levels of automatisation. The use of CAPI for verification and improvement of the quality of the photointerpretation is highly recommended [2]. Table 2 summarises some of the important characteristics of the database.

TABLE 2. Basic characteristics of the CORINE Land Cover Database

Area coverage: 3.9 million km^2, 24 countries (1999) fully completed

Method: computer assisted satellite photointerpretation

Satellite images: LANDSAT, SPOT, etc.

Ancillary data: topographic maps, vegetation maps, land use and land cover statistics, etc

Working scale: 1:100 000

Minimum mapping unit: 25 hectares

Minimum linear feature width: 100 m

Nomenclature: hierarchically structured in 3 levels: 5 classes for level-1, 15 classes for level-2 and 44 classes for level 3

Cartographic projection: Lambert azimuthal equal area

Various projections in country databases

Implementation: national teams

Supervision: Land Cover Technical Unit

Overall classification reliability: better than 85%

Geometric accuracy compared to topographic map: 100 meters RMS or better

Mean cost: 5 ECU/km^2

2.2. NOMENCLATURE

TABLE 3. CORINE Land Cover nomenclature [8]

LEVEL 1	LEVEL 2	LEVEL 3
1. ARTIFICIAL SURFACES	1.1. Urban fabric	1.1.1. Continuous urban fabric
		1.1.2. Discontinuous urban fabric
	1.2. Industrial, commercial and transport units	1.2.1. Industrial or commercial units
		1.2.2. Road and rail networks and associated land
		1.2.3. Port areas
		1.2.4. Airports
	1.3. Mine, dump and construction sites	1.3.1. Mineral extraction sites
		1.3.2. Dump sites
		1.3.3. Construction sites
	1.4. Artificial, non-agricultural vegetated areas	1.4.1. Green urban areas
		1.4.2. Port and leisure facilities
2. AGRICULTURAL AREAS	2.1. Arable land	2.1.1. Non-irrigated arable land
		2.1.2. Permanently irrigated land
		2.1.3. Rice fields
	2.2. Permanent crops	2.2.1. Vineyards
		2.2.2. Fruit trees and berry plantations
		2.2.3. Olive groves
	2.3. Pastures	2.3.1. Pastures
	2.4. Heterogeneous agricultural areas	2.4.1. Annual crops associated with permanent crops
		2.4.2. Complex cultivation patterns
		2.4.3. Land principally occupied by agriculture, with significant areas of natural vegetation
		2.4.4. Agro-forestry areas
3. FOREST AND SEMI-NATURAL AREAS	3.1. Forests	3.1.1. Broad-leaved forest
		3.1.2. Coniferous forest
		3.1.3. Mixed forest
	3.2. Scrub and/or herbaceous associations	3.2.1. Natural grassland
		3.2.2. Moors and heathland
		3.2.3. Sclerophyllous vegetation
		3.2.4. Transitional woodland-scrub
	3.3. Open spaces with little or no vegetation	3.3.1. Beaches, dunes, sands
		3.3.2. Bare rocks
		3.3.3. Sparsely vegetated areas
		3.3.4. Burnt areas
		3.3.5. Glaciers and perpetual snow
4. WETLANDS	4.1. Inland wetlands	4.1.1. Inland marshes
		4.1.2. Peat bogs
	4.2. Marine wetlands	4.2.1. Salt marshes
		4.2.2. Salines
		4.2.3. Intertidal flats
5. WATER BODIES	5.1. Inland waters	5.1.1. Water courses
		5.1.2. Water bodies
	5.2. Marine waters	5.2.1. Coastal lagoons
		5.2.2. Estuaries
		5.2.3. Sea and ocean

The standard CORINE Land Cover nomenclature includes 44 land cover classes (Table 3). These are grouped in a three-level hierarchy. The five level-one categories are: 1) artificial surfaces, 2) agricultural areas 3) forest and semi-natural areas, 4) wetlands, 5) water bodies. All national teams have to adapt the nomenclature according to their local landscape conditions. For national purposes it is allowed to subdivide further any of level-3 elements of the nomenclature. E.g. two level-4 categories were used in Ireland to characterise pastures (2.3.1) of different quality [11]. Similarly, two different types of inland marshes (4.1.1) and peat bogs (4.1.2) were defined in Estonia [1]. The European database however includes only level-3 categories.

Special features of the nomenclature are the categories of "Heterogeneous agricultural areas". They are formed by objects, (e.g. plots of arable land, areas of natural vegetation, etc.) which alone would be smaller than the minimum mapping unit (25 hectares). E.g. category 2.4.2 have been introduced to characterise mixed agricultural areas: mixtures of any two of the following cover types: arable land, pastures, vineyards, fruit trees and berry plantations. Class 2.4.3 is to characterise agricultural areas with significant amount of natural formations (e.g. patches of forests, areas of scrub, grasslands, wetlands or water bodies). These are very useful tools to characterise a heterogeneous landscape at scale 1:100 000.

3. Land Cover Mapping at Scale 1:50.000

Standard CORINE Land Cover data are especially useful at European level. To satisfy regional or local needs better, more details are needed both in terms of geometry as well as in thematic content. Several initiatives exist to extend CORINE Land Cover methodology into working scale of 1:50 000 and even 1:20 000 [6]. In the frame of the Phare Programme an experimental project has been executed at the scale of 1:50 000 including 120 map sheets in four countries: Czech Republic, Hungary, Poland and Slovak Republic. It was possible to use 4 hectares as minimum mapping unit using integrated SPOT PAN and Landsat TM data. An international team of experts has extended the standard nomenclature with level-4 classes representing the landscape conditions of the above four countries [9]. The number of level 4 classes was about twice of level 3 ones.

The experimental project proved the possibility of the CORINE Land Cover Mapping at larger scale. One of the activities of PTL/LC has been to further extend the level-4 nomenclature including all Phare countries. New version of the CORINE Land Cover nomenclature at scale 1:50 000 for Phare countries has been finished in 1998. This nomenclature includes 104 land cover classes and could be a base of an all-European level-4 nomenclature. In 1999 Hungary has started a national CORINE Land Cover Mapping at scale 1:50 000 based on SPOT-4 imagery, to support the needs of the Ministry of Agriculture and Regional Development and the Ministry of Environment.

Main benefits of the extended nomenclature and the 4 hectare minimum mapping unit compared to the standard CORINE Land Cover Mapping are:
- much more thematic detail in the "artificial surfaces" group which has the strongest impact on the environment,
- agricultural categories support better agrostatistics and the needs of habitat mapping,
- more discrimination in forests and semi-natural vegetation and in wetlands, which are important for nature conservation and biotope mapping,

decreased percentages of heterogeneous agricultural classes using smaller minimum mapping units.

4. Updating and Change Detection

Updating is a central question of any databases including features, which change in time. The CORINE Land Cover database can fulfil its aims only if the database is regularly updated. The proposed average updating frequency of the CORINE Land Cover database is 10 years. This does not mean that changes cannot be faster in certain areas (e.g. urbanization). Having land cover data for more than one date, one has a possibility to analyze land cover changes and to make predictions for the future.

4.1. UPDATING

CORINE Land Cover Mapping is a human labour-intensive methodology, requiring skilled photointerpreters. Because of the nature of nomenclature and the rules of interpretation, updating also cannot be automatic. Due to the fact, however, that land cover changes are generally slow, there is no need to repeat the interpretation in the course of updating, only to recognise changes which have happened from one date to the other. Having a proper computer support, this process is evident for a photointerpreter, familiar with the CORINE methodology. Therefore updating is significantly cheaper than producing the basic database.

The updating process is based on the computer-assisted photo-interpretation (CAPI) technology, with simultaneous use of the basic CORINE Land Cover Map, the corresponding satellite image map, and the new satellite image map. Most important features of the necessary CAPI software are: raster background handling capabilities, geographically linked multi-window environment, ability to edit different databases in different windows, building-up and checking of the database topology and general image processing capabilities. In addition to several commercial GIS/IP processing softwares that support this list, JRC has developed the Co-PILOT (CORINE Photo-Interpretation Land Cover Oriented Tool) software, which includes additional, specific CORINE related features [12]. The updating procedure usually reveals errors in the original database that first should be corrected, in order to avoid detection of false changes [3, 14].

There are plans to update European CORINE Land Cover data using a satellite image snap shot of Europe for the year 2000, referred to as IMAGE2000. It will be a multi-purpose image archive, which will be used for several environmental applications. The full update referred to as CLC2000 should be finalised by 2003 [16].

4.2. EVALUATION OF CHANGES

Once we have produced the CORINE Land Cover database for the dates T_1 and T_2, change detection is an automatic procedure. The change database includes polygons with attributes related to T_1 and T_2. The change database can be visualised by printing evolution maps and summarised using statistical tools. The evolution matrix (contingency table) is the most detailed statistical descriptor of summarised area changes between the two dates. Its diagonal elements represent areas of no change, while off-diagonal elements relate to area changes between T_1 and T_2. Having 44 level-3 categories, the maximum size of evolution matrix is 44*44. (In practice the evolution matrix includes lots of null values, because of impossible transitions between several category pairs). Summary statistics for T_1 and T_2, the areal change for each category and the total change can be derived from the evolution matrix.

There are some other useful indicators of changes [7]:

- The normalised relative area change answers the question: which are the classes with the largest area increase or decrease per year?
- The relative occurrence change answers the question: which are the most dynamic classes relative to their original frequency of occurrences?
- The relative updating frequency (the ratio of the proportion of modifications and the proportion of area of a given class) answers the question: which classes are updated more than the average and which less then the average?

These indices can be used to compare changes derived for areas of different size, and/or different time span and to work out operational updating scenarios.

5. Use of CORINE Land Cover Data

The ETC/LC organised in 1997 a Workshop on land cover applications at European scale. The workshop illustrated how CORINE Land Cover data are actually used by compiling 28 applications from projects from all parts of Europe to exemplify specific domains or application fields. The objective was to demonstrate and assess the land cover data in environmental and integrated applications particular reference to needs of the EC environmental policies. Table 4 gives a survey of the environmental domains illustrated and the regional coverage of the examples.

TABLE 4. Application domains

Application field	Coverage of examples
Nature conservation	European
Water management	Regional
Forest fragmentation	European-Regional
Coastal management	European
Transport	European
Air pollution distribution	Regional-European
Agriculture	Regional
Urbanisation	Local comparison of European cities
Structural funds/Land planning	Regional
Soil degradation — desertification	Local-Regional
Hazards (forest fires, flooding)	Regional

The examples from both EEA member states and Phare countries illustrate the conclusion: CORINE Land Cover data, when combined or integrated with other data sets in a GIS environment, constitute georeferenced data of basic importance to environmental analysis, evolution studies, evaluation of pressures and trend analysis, related to spatial problems or issues. More information is accessible on the Website as an annex to the Proceedings from the 1997 Workshop, under the heading 'Workshop on Land Cover Applications' (http://www.mdc.kiruna.se/etc/ Workshop/contents.htm).

6. Conclusions

Results of the CORINE Land Cover Project have been integrated into the environmental database of the European Union. Due to the clearly defined, relatively simple technology and good project management, an up-to-date, uniform and harmonised land cover database has already been compiled for a large part of the continent. Its main aim is to provide a quantitative basis for defining the environmental policy in Europe. The database is also on national levels widely used and fosters multi-country co-operation on various problems related to the environment. High-resolution satellite imagery and advanced methods of spatial data processing have been the two most important technical catalysts of these achievements. The same tools make it possible to update and "down date" the database, hence providing information on land cover changes. According to extended experiments, the methodology of the CORINE Land Cover can be applied at larger scales as well to provide more spatial and thematic information for regional decision-making.

7. References

1. Aaviksoo, K.(1997) Personal communication.
2. Büttner, G. (1997) Computer assisted verification and correction of CORINE Land Cover photointerpretation, *Phare CORINE/EEA Newsletter* 6, 8-11.
3. Büttner, G., Maucha, G., Bíró M.(1998) Land Cover change detection using the CORINE methodology. *Resources and Environmental Monitoring, ISPRS*
4. Commission VII Symposium, Budapest 1998 (International Archives of Photogrammetry and Remote Sensing, Vol. XXXII Part 7), 685-690.
5. Dewos, W.(1997) The first CORINE Phare inventories CD-ROM is ready. *Phare CORINE/EEA Newsletter* 5, 18.
6. ETL/LC (1997) CORINE Land Cover Directory. Prepared for the European Environment Agency.
7. ETC/LC (1997) Assessment of the existing experiences of the 4th and 5th level CORINE Land Cover nomenclature. Prepared for the European Environment Agency.
8. ETL/LC (1997) Report on updating. Prepared for the European Environment Agency.
9. European Commission (1993) CORINE Land Cover, Technical Guide, EUR12585, Brussels, Luxembourg.
10. Feranec, J., Ot'ahel', J., Pravda, J .(1995) Proposal for a methodology and nomenclature scale 1:50.000 CORINE Land Cover Project, Final Report, Institute of Geography, Slovak Academy of Sciences, Bratislava.
11. Jaakkola, O. (1994) Finnish CORINE Land Cover – A feasibility study of automatic generalization and data quality assessment. Reports of the Finnish Geodetic Institute.
12. O'Sullivan, G. (1992) CORINE Land Cover Project (Ireland), *Survey Ireland*, **November 1992,** 7-43.
13. Perdigao, V., Annoni, A. (1997) Technical and methodological guide for updating CORINE Land Cover Data Base, JRC/EEA Report.
14. Phare (1996) Phare Natural Resources CD-ROM, Phare Programme, European Commission DG IA B5.

15. PTL/LC (1998) Methods of Computer-Assisted Photo-interpretation and Land Cover Changes Detection. Prepared for the European Environment Agency.
16. Steenmans, C. and Bossard, M. (1996) CORINE Land Cover data inventory for the first six Phare countries now complete. *Phare CORINE/EEA Newsletter* **4**, 7-10.
17. Steenmans, C.(1999) European Topic Centre on Land Cover, EUROSTAT Working Document, Working Party "Land Use Statistics" of the Agricultural Statistics Committee EEA, 31-May- 1 June 1999, Luxembourg
18. 16. Swedish Space Corporation (1994) CORINE landtackning – ett pilotprojekt i Sverige. Technical Report in Swedish, with English summary.

COMPUTER-ASSISTED LARGE AREA LAND USE CLASSIFICATIONS WITH OPTICAL REMOTE SENSING

NIKOLAS PRECHTEL
Institute for Cartography
Dresden University of Technology
Mommsenstr. 13
D-01062 Dresden
Germany
prechtel@karst8.geo.tu-dresden.de

Abstract

An overview shall be given on operational methods and steps involved, when optical remote sensing data shall be digitally processed to result in a land use data base, which certainly forms one of the most prominent tasks of remote sensing. Questions of terminology (especially land use and land cover) will be covered, as well as data selection and acquisition, noise correction, geo-coding, classification, post-processing and map production. Obviously, only guide-lines can be given and it would be ways beyond the scope of this article to cover the whole spectrum of interesting approaches. It must be pointed out, that high quality demands call for an adequate regard of ancillary data; their use is still hampered by technical barriers as disperse storage and solely analogue availability, various geometric projections, and others. Moreover, commercial image processing software for use with remote sensing data does hardly provide any tools to imbed a-priory knowledge. Geo-scientific knowledge on vegetation patterns, crop-rotation systems in agriculture and phenological information around the time of image acquisition ('dynamic vegetation models') can significantly improve the classification results. The core task is the design of an efficient classification method, which must be carefully adapted to the class specifications. A 'brute-force' approach aiming at results in a single step with one universal classifier cannot be recommended. Examples for a more sophisticated solution are basically taken from a large-area project for the state of Saxony (Germany): A combination of default functions and additional procedures was allowing to profit from a selective choice of spectral bands, classifiers and (post-)processing steps at every node of a hierarchical classification tree. Wherever local image features were performing insufficiently, textural or form attributes have been included. The cited project was accompanied by the generation of a set of 15 land-use maps in a standardised layout. Finally, some remarks will be given concerning a potential project for a comprehensive land-use map in a less-developed area like Albania.

M.F. Buchroithner (ed.),
Remote Sensing for Environmental Data in Albania: A Stragegy for Integrated Management, 101–126.
© 2000 *Kluwer Academic Publishers. Printed in the Netherlands.*

1. Introduction

1.1. IMPORTANCE OF LAND USE / LAND COVER DATA

Land use and land cover changes are undisputed aspects of environmental change at any spatial level, from local to global. Under steady-state physical conditions land cover patterns are changing by reacting on slow, but long-lasting processes (e.g. succession from wetland vegetation to a closed forest stand in a delta in phase with soil development and increase of distance between surface and ground water table) or on spatially selective, high-energy short-term impacts (e.g. mass movement or forest fire causing a sudden breakdown of a forest stand). Human activities, however, have terminated steady-state conditions more or less all over the world within the last centuries and became the main driving force behind land cover change; this is partly due to the dynamics in the human societies (population dynamics, division of labour, economic level, etc.) or due to reactions on the often unintended alterations of the physical environment (as soil erosion, soil contamination, water shortage, etc.). Hence, human activities are imbedded into an environmental control circuit while occupying an active as well as an reactive position (comp. Figure 1). Knowledge on these dynamics, not only in an abstract, but precisely geographically located form, is crucial to feed environmental models and to help in tackling observed problems by scientifically funded action.

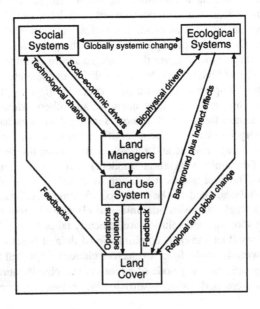

Figure 1. Fundamental structure of land use and land cover change. Land managers (the farmers, loggers, etc.) are influenced by both the social and ecological systems in which they operate. Their activities, such as clearing forests, burning savannahs, or building terraced paddies, constitute a specific land use system, which in turn changes natural land cover (figure taken from [4]) .

1.2. DEFINITIONS

At first, one should look closely at the terms land use classification and land cover classification:

Land use classification, to our understanding, means the assignment of basically functional, in a time scale of years persistent, and in their physical impact on the environment dominant human utilisation classes (e.g. arable land) to a contiguous area of the upper boundary layer of the geo-sphere.

Land cover classification, on the other hand, means a class assignment according to (predominately seasonally) varying material and structural states of the boundary layer (e.g. bare, ploughed soil). There are, of course, areas, which are subject to perpetual variations of this material state like low-lying river banks or tidal sea shores (flooding).

Optical sensors are mapping the spectral remission of solar radiation from the surface towards the sensor in the visible (VIS) and infrared part of the electromagnetic spectrum. A further division of the reflective infrared (0.67μm – 3.0μm) into a near infrared (NIR: 0.67μm – 1.5μm) and short-wave infrared part (SWIR: 1.5μm – 3.0μm) is commonly used ([1] p. 215). The basic information of optical remote sensing, to be retrieved from this spectral signal, is the material and structural composition of the surface, which is, in the easiest case, uniform for a ground segment as a real-world equivalent of one image pixel. The inference of land use from land cover information is, at least theoretically, an independent following process, which requires stable relations between the imaged material state of one ground segment (or a characteristic pattern) and a land use type [2]. Because of the temporal variations (comp. above) it will often be necessary to secure that typical conditions (e.g. water lines in flat land, comp. e.g. [3], chapter 2.4) are prevailing at image acquisition time to get valid and unbiased classification results.

A characteristic of operational *large-area classifications*, to be discussed in the present paper, is the demand for more than one scene to obtain full coverage. This implies different dates of data take and, therefore, different imaging and ground conditions (state of atmosphere, phenology), while in the same time the results have to be (completely) scene-independent. With the ubiquity of high-performance processors and storage media, the tackling of a large data amount is now subordinate to the problem of two or more scene-dependent image physiognomies of the same land use class.

1.3. REMARKS ON THE INTERACTION OF GIS AND IMAGE PROCESSING

At nearly all stages of a land use / land cover classification project GIS can and should be integrated. A main use can be seen in selecting, pre-processing and export of ancillary data to support and evaluate the digital classification. A summary is given in Table 1.

Despite of potential benefits, a direct link between GIS and image processing software, enabling classifiers to work on hybrid geo-data structures, has not fully been introduced in commercially available software packages.

TABLE 1. Synergy of digital image processing and GIS-tasks.

(Raster-) Image Processing Task	linked by ... to	GIS Task (basically vector data)
Image data analysis (header information, meta-data, image statistics)	Demand on external geo-data	Semantic and spatial selection of digital ancillary geo-data
Implementation of classification method	Structure of classification key	Filtering and task-specific upgrading of selected ancillary data
Rectification, geo-coding	Demand for common geometry of all geo-data involved	Common geodetic referencing / transformation of all data sets
Classification and generalisation of primary results	Evaluation and generalisation	Combination with reference data, verification by statistical methods, generalisation of classification results
Set-up of raster data colour scheme	Map layout	Map composition, compilation and symbolisation of added topographic data
Unspecific	Import / export and format conversions	Unspecific

2. Data Acquisition

2.1. PRE-ACQUISITION CONSIDERATIONS

A careful selection of image data for the digital classification must be carried out. For convenience and efficiency, one should aim at a complete coverage of the project area by imagery of one or more selected sensors. Thus, multi-sensorial classification does not indicate the use of one sensor in one part of the study area and of another sensor in the complementary part. In spite of some convergence in spectral band design of operational sensors (comp. e.g. [5], [6]), there might be preferences, especially for the short-wave infrared information not 'visible' for all sensors. A deficiency in the spectral domain of a single image might be compensated by additional data of the same sensor, but from a different season (*multi-temporal*), or by an improvement of the spectral resolution through a *multi-sensorial* approach. Looking at time efficiency, a multi-temporal approach with data from only one sensor might be superior, since similar or identical geometric and sensor calibration models can be used. Possible selection criteria might also be in a mutual conflict. (comp. Table 2).

2.2. INFORMATION SUPPORTING A RATIONAL DATA SELECTION

Among various useful information, only a few can be pointed out. They are related to the most variable components imaged by an optical sensor, atmosphere and vegetation.

TABLE 2. Criteria and potential conflicts in selecting an optimised image data set for digital land use classification.

	General Objectives		Exemplary Criteria Catalogue	Closely related to
1	Low data costs	1.1	Cost/Coverage-Ratio	
		1.2	Scene location over project area	3
		1.3	Archived image or special acquisition	6,7,8
2	Detailed classification key / high accuracy	2.1	Mono- or multitemporal classification	3,4.2,5,6,7,8
		2.2	Mono- or multisensoral classification	4.1,4.2
3	Most suitable orbit configuration for classification (sensor preference)	3.1	Temporal resolution / repetition	2.1,5,6, 7,8
4	Most suitable sensor design for classification (sensor preference)	4.1	Spatial resolution	2.2
		4.2	Spectral band layout	2.1,2.2
5	Best phenological period(s) for classification	5.1	Agriculture: crop differentiation	2.1,3,6,7,8
		5.2	Phenology of forest trees	
6	Avoidance of 'untypical' land cover configuration at time of data capture	6.1	Actual waterlines (seashore, lakes, rivers)	1.3,2.1,3,5,8
		6.2	Recent hazards with major influence on land cover	
		6.3	Snow cover	
7	Minimum cloud cover in imagery	7.1	Percentage, type and distribution	1.3,2.1,3,5,8
8	Importance of actuality for classification objective	8.1	Speed and distribution of land cover change	1.3,2.1,3, 5,6,7

Widely used are:

- Meteorological data on average and daily cloudiness and on phenology;
- Data on farm and forest management (especially cultivated crops and rotation systems).

Concerning mean values, meteorological tables in digital form or handbooks have been published. They allow - in combination with orbit data - an estimation of the probability to obtain images with low or no cloud cover [7]. A factual cloud cover of archive images in coarse percentage classes is an item in the catalogues, but no information on the cloud distribution is provided, which might eventually be irrelevant for the study area. That leads over to a quality assessment of archive images using so-called quick looks (usually a NIR-band in reduced resolution), but a careful inspection should imply a check of meteorological data like meteorological visibility, water vapour, and METEOSAT image data to account for thin haze and smog, often invisible in a quick-look generated from the near-infrared channel [8].

Meteorological statistics will normally contain *phenological information* from long recordings. With (basic) knowledge about the vegetation in the project area, phenological data greatly helps in modelling the physical appearance of a landscape at a certain date in the seasonal cycle (green-wave over forest and agricultural land, work calendar in the field, maturity stages and harvest dates of important crops, etc., comp. [9]). Since mean occurrence dates alone might have a limited value, they must be augmented with actual information, which is to some extend published by the national meteorological surveys (e.g. 'Agrarmeteorologischer Wochenhinweis' = Agrometeorology Weekly of German Weather Survey) and, for example in the German case, partly accessible through the internet [10]. For the classification job an ideal image will show a seasonal state of a landscape, where the relevant land cover classes are characterised by maximum mutual physiognomic difference and a minimum internal variance. This goes most likely along with specific spectral signatures. But still we will face the question "How unique are spectral signatures?" [11]. In spite of a systematic treatment of the seasonal influence on a discrimination of vegetated surfaces some 40 years ago using aerial photographs [12], we must still state a lack of long-term *systematic spectral field measurements* for vegetated surfaces (comp. [13]). This does not mean to ignore an extensive list of project-specific hyperspectral plant reflection libraries, reflecting decades of laboratory and in situ measurements (only links for further reading can be given, e.g. [14], [15]). A internet-based digital library of standardised vegetation spectra taken in short intervals over the whole vegetative cycle would greatly assist the remote sensing community in data selection and classification, and would form an ideal complementary to the meteorological information. So far, we are confined to task-specific spectral measurements with a limited operational value. A standardisation of phenological stages is already on its way [16].

A lot more assistance exists for the actual image selection. Coverage patterns can be derived by specific software for most of the operational sensors (e.g. Display Earth Remote Sensing Swath Coverage for Windows [17]), a prerequisite for calculation of costs, special request orders, and, most important, for the timing of synchronous field observations. Archived data can comfortably be checked via internet, and search machines enable access to distributed data-banks (comp. [18]).

2.3. DATA ORDERING

The providers offer data at different *processing levels*, from raw over system-corrected to sometimes geo-coded image data. A conflict occurs between a maximum control on each processing step which can only be achieved with raw data, and time-economy calling for higher processing levels. A systematic treatment of all initial geometric corrections is given by [19]. In most operational projects data with a standard processing level will be acquired. In that case system corrections accounting for orbit influence on geometry, earth rotation at overpass and basic radiometric correction have already taken place. The user has to care for further pre-processing as noise removal, rectification and geo-coding, and, eventually, radiometric calibration. High-precision rectifications as further processing step, based on common geodetic reference material and Digital Elevation Models (DEM), are also partly offered by the image providers at higher costs.

3. Noise Removal

Potential disturbance of image information can usually be attributed to failures or problems in the imaging process, in the data down-link, in the archiving at the ground station, in the pre-processing of the data by the provider, or in the transfer of digital data from the provider to the user. This can result in:

- Line drop-outs,
- Line duplications,
- Punctual grey-value blunders,
- Striping of imagery due to inhomogenities in the sensor calibration.

Since most of the occurring problems are related to the sensor physics, the easiest detection and restoration is applied to the *original image matrix*. This means, that an original scan line of the instrument still forms an image line. Correction should precede all further geometric and radiometric processing of the image data. When only one spectral band is disturbed by locally missing information, the correlation matrix of all spectral bands of the image can be used for restoration. Linear or punctual drop-outs in all bands might find an appropriate correction by grey value estimations based on the neighbourhood using auto-correlation functions. Image striping will severely affect a classification task; it will weaken the selectivity of classes, and deteriorate especially texture measures generated from small windows. In case of a periodic character, the noise in the satellite image (e.g. [20]) can be eliminated by a virtual subdivision of the image and a following calculation of histograms; then, the grey-values will be modified in such a way, that the mean value and dynamic range of all subsets will be adjusted [21]. A similar result can be achieved by an application of the Fourier Transform, masking of the problematic frequency in the power spectrum, and subsequent retransformation. In case of non-periodic noise, carefully designed filters might be applied, however, at the expense of some unintended grey value modifications all over the image. An efficient 'destriping' method, without detrimental effects on the undisturbed data and also practicable with geometrically corrected data, is based on multi-step morphological filtering to create a precise localisation of the stripes, and subsequent grey-value adjustment [22].

4. Atmospheric and Topographic Correction

The objective of the corrections is to derive *spectral reflectance values* of the ground elements (pixels) from grey values. The latter shall be freed from signal contributions not related to the local interaction between solar radiation and the surface, our observation target. This is associated with the suppression of the following effects by calculating their influence on the signal:

- multiple scattering of light in the atmosphere,
- uneven solar irradiance by varying slope, aspect, elevation and state of atmosphere over the area covered by the image(s),
- adjacency effects (mutual influence of neighbouring grey values in case of high reflectance contrasts).

These corrections become in particular necessary, when 'multi'-data sets are exploited to elucidate reflectance properties. Material (e.g. phenological) change or stability in time is an important discrimination feature in case of two or more data takes (best from the same sensor). With the help of a roughly synchronised view by two different sensors on the same spot, the individual lay-out of spectral bands will provide an image stack with increased spectral resolution.

For a classification strategy, which is reflected by Hill [23] in this issue, the calculation of reflectance is a prerequisite: the spectral unmixing, which tries to describe the surface of a ground segment out of a set of spectrally defined endmembers, for which abundances (totalling in 100%) are calculated.

In general, there is no fully satisfying physical solution to get 'true' reflectance values for a complex terrain mapped by a satellite sensor. This is due to a bundle of necessarily simplified assumptions, especially, in the scattering characteristics of the a-priori unknown surface, the optical depth of atmosphere constituents in their spatial variation, and the roughness (or 3-dimensionality) of the scattering surface in its influence on the radiation transfer. Valuable existing solutions try to make use of iterative corrections based on atmosphere (average conditions in climatic zones during a specific season) and surface properties. Operational software assistance is given; within the ATKOR software a spatially adaptive atmospheric correction can be calculated for most of the operational digital space sensors [24]. A basic idea of image-based solutions is the inference of a spectral sky-light fraction from a so-called tasselled-cap transformed image [8]. As a final correction step, also the adjacency effect can be taken into account.

Topographic correction tries to subdue the relief influence on incoming solar irradiance (flux density and geometry). It is basically controlled by the angle between the normal vector of a ground element and the solar beam radiation vector; correctly, the view angle of the sensor must be included (bi-directional reflectance distribution function, BRDF). Unfortunately, various surface structures (e.g. roughness) and geometries (e.g. needles or broad leaves) are complicating an uniform correction by an individual scattering geometry. In the end, an image of mountainous terrain with high dynamics on the sun-exposed slopes and low dynamics on shady slopes shall be optically flattened, so a correlation between aspect and reflectance of a given surface tends towards zero. In this step one must keep in mind, that, looking globally at the image, a correlation is partly due to different land cover on sunny and shady slopes. As in the case of atmospheric correction, there is software assistance, which allows to generate a synthetic irradiance model for a given DEM and sun position for the subsequent correction of the image data. The quality of the result is limited by three main factors: accuracy and resolution of the DEM used, mutual geometric fit of image and DEM, and correctness of the scattering model for a surface type. Correction artefacts by bad fit or inadequate quality of the DEM can quite easily be detected along ridge lines in rugged terrain.

5. Geo-coding

No extensive technical treatment of this topic shall be given in this context; geometric and geodetic aspects are part of every remote sensing teaching book, and also partly discussed in this issue [25]. However, within the present application-centred context it must be stressed, that a reliable geometry is a prerequisite for:

- the integration of all digital ancillary data in the classification and map generation,
- the use of the data in a GIS-environment.

For the decision about an appropriate rectification method, the geometric accuracy demands should be clarified first. In any case, the accuracy limits are given by the spot identification accuracy in the image as a function of the sensor's ground resolution and the local contrast, and the accuracy of the reference source, in many cases a map. It is quite clear, that for the ground control point selection *only temporally invariant structures* should be chosen. An edge in the waterline formed by a jetty-wall might be a perfect reference object, while a waterline on a tidal beach might be the opposite. For the sake of geometric reliability, a large number of even-distributed control points, forming a tie between image and reference source, is desirable. A proper quality check also implies the use of *independent check points*, which are not controlling the rectification process via transformation matrix. Whether a parametric method with a higher operator and calculation time is necessary or not, can be determined by the relief displacement as a function of off-nadir view angle and elevation variation (comp. [26]). Whenever image sets will be rectified individually instead of blockwise by bundle adjustment, a careful check of the seam lines between the rectified images for geometric consistency is required. A further qualitative and quantitative control step makes use of residual vectors to be calculated for all ground control points. Residual vectors are connecting measured target locations with the new location in the geometrically transformed image matrix. The results must not show any directional bias and the frequency distribution of the vector lengths should decrease monotonously. For nadir images of the operational optical sensors, the standard deviation of the residuals in row- and column-direction can normally be kept below a pixel equivalent.

6. Image Subdivision and File Name Conventions

6.1. IMAGE SUBDIVISION

In any large-area classification project the extraction of semantic information will be carried out in more than one step. This is due to the use of imagery from different dates, or even dates and sensors, which store the required information in a different way. A further division of individuals scenes into subsets for the classification can be favourable:

- to reduce machine answering time in extensive calculations,
- to facilitate the operator's overview on the quality of (intermediate) results, and
- to 'slim' the feature space via reduction of land cover variety, making use of predefined complex geographic landscape units,
- to introduce prior knowledge on class distributions or mixing proportions into the classification.

If a subdivision seams appropriate, and the cutting lines will be delineated according to geographic landscape units, all plausibility checks of (intermediate) classification steps like occurrence of certain classes and corresponding frequencies can be based on the prior knowledge about the characteristics of these units. Spatial stratification and individual classification is well established, but labour-intensive [27]. A further disadvantage of long, irregularly curved cutting lines is obviously an additional effort to secure consistency in the class assignment after re-assembly of the subsets. In the frequent situation, that topographic maps will form the principal reference source, and no digital landscape maps in the right scale are accessible, a schematic subdivision according to the map sheets might be the best choice. In the cited Saxony land use classification project, the pattern of the official German 1:100 000 topographic map series has been the base, leading to subsets of 7-band Thematic-Mapper data with around $3 \cdot 10^6$ pixels.

6.2. FILE NAME CONVENTIONS

The task of defining binding conventions for a file nomenclature is leading over from the pre-processing to the classification. Basically, the file names have to contain meta-information on geographic location of the data, processing level and history, and data type in a most compact and, therefore, coded form. From these demands it becomes clear, that the system must be set up along with the classification method (e.g. Table 3). File name conventions are crucial in any extensive digital processing of geo-data to avoid a mess on the hard disk, to reduce documentation time, and, thus, to smooth the co-operation of the project team in charge. A dependable name and file system should, nevertheless, be accompanied by a *logbook*, which is normally a default feature of an image processing software, and only needs minor editing and commenting during and at the end of each session.

TABLE 3. Example for file name conventions in a land use classification project.

Exemplary File Name	Parts of Code		Explanation
	Digit 1	D	Scene code (after major town Dresden)
	Digit 2	3	Subset code (sequential subset number)
D3_C3Fi1.img	Digit 3	_	Separator between location code and processing code
	Digit 4	C	Type of processing: C for Classification
	Digit 5 / 6	3f	Processing level according to classification tree, indicating level 3, and processing class f for forest
	Digit 7 / 8	i1	Process history: first in a sequence of several steps (iteration 1)
	Digit 9	.	Separator between file name and extension
	Extension (Digits 10-12)	img	Extension indicating data type (img for image)

7. Classification

7.1. STRUCTURE OF THE CLASSIFICATION KEY

As pointed out in chapter 1.1, a land use / land cover classification is multi-purpose data, which, theoretically, might serve any agency or scientist being concerned with modelling and planning of our natural and man-made environment. The potential of the remote sensing data as well as the methodological know-how of the interpreter(s), determine the limitations of semantic and spatial detail of the output. This compares to the scale problem in cartography, and, similarly, tells about the data suitability for a given task. Unsurprisingly, the classification key will often reflect a compromise between the interest of *different user groups* and the *classification potential* of the information source. Since a standard task is discussed, there are many existing examples and there is rarely a need for complete re-design. Already in early examples, we find tree-shaped (hierarchical) classification trees [28]. This is an obvious and logical structure for a classification task, leaving room for refinements at the upper end without altering the basic architecture, and enabling the interpreter of remote sensing data to define realistic targets. These targets might all be located in one level, but more often we will have to draw a curved line between classifiable and not classifiable parts across the classification tree. This dividing line is data-, operator-, and technology-dependent.

7.2. CLASS DEFINITION STANDARDS

A workable classification key must contain precise class definitions. Several major problems are apparent:

1) The *number of classes* must be limited, but, at the same time, the total range of land cover or land use of a large territory, has to fit into the system.
2) A decision has to be made, if a class is defined according to *functional* (land use) *or material principles* (land cover). Without explicit reference, many classifications mix these principles, what might be exemplified:
   The class 'arable land' refers to the use, but shows manifold appearance on ground and in images throughout the year. It will be formed as an union class of several surface states with image features, which might not allow a more specific class inference (into winter wheat, rye, maize, etc.). The (super-)class 'forest', derived from satellite imagery, will contain land with a minimum ground cover by trees, a structural definition in contrast to the functional one in a topographic map. All parcels in a state between tree felling and reforestation close to crown closure will not show up as forests. Herbaceous understorey vegetation and brushwood are dominating the reflecting surface. As in many other cases, the image gives no indication, whether a surface is in a transitional or in temporally stable state.
   Summarising it can be stated, that a high modelling potential for later users is given, when at least a rough inference is possible from the classification to (average) surface characteristics at any time of the year (comp. Figure 2). That will normally require additional data to account for predictable (seasonal) dynamics. For that reason, a small patch in the middle of a suburban residential area, classified as

forest, might be functionally incorrect (since it is, for example, part of a large garden site), but allows in the same time much more insight in the structure than a homogeneous patch classified low-density built-up land.

3) The problem of *transitional states* of land is a very frequent one as is the problem with a *lack of sharp boundaries* between land cover units: how to classify barren land being in various states of succession from a former pasture over brush land to a pioneer forest? Similar problems turn up with disused quarries and mining sites in various stages of re-cultivation or natural succession.

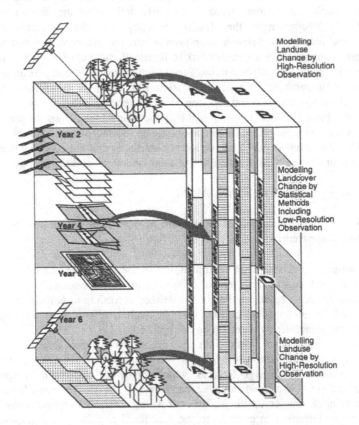

Figure 2. Modelling of continuous land cover change by combination of high-resolution satellite observation in long cycles (e.g. 6 years) and statistical prediction using meteorological satellite-data, statistics and maps to account for short-term variations. Cells A, B, C, and D indicating observation/modelling cells. Segments in the columns marking major changes of land cover, partly predictable (seasonal vegetation change), partly unpredictable (transformation from forest to built-up).

7.3. MIXED INFORMATION SOURCES

From a user's point of view land cover data might also be extracted from mixed data sets, for example from remote sensing and *topographic data*, if the overall quality (level of detail and reliability) can be increased. Such an integration of external data can be sensible, especially when linear features in the landscape (such as small rivers or major road network) shall be realistically embodied. 'Under-sampled' landscape features with a typical width in the range or slightly below the pixel resolution can, if ever, only be classified in fragments; this must obviously lead to biased final statistics of class distribution and percentages. It is however crucial, that objects from external data bases are stable in location and distribution (as rivers in humid climate), to assure a consistent actuality after integration. Also all external input should be marked in meta-data, respectively in the map legend. These aspects are quite often disregarded [29].

7.4. CLASSIFICATION METHOD

A classification method will, in any case, be developed in a task-specific way. Furthermore, the available software functions, the time budget, and ancillary data often imply methodological restrictions. Nevertheless, basic principles can be named, which seem to be universally valid:

- Minimisation of user interaction, or, vice versa, maximisation of procedurally computed results;
- Use of the dominant feature or feature combination (as spectral signal, form, size, texture, orientation, association), with dominance being defined according to the effect of the specific feature space selection on separability measures;
- Preference of simple, fast-working and spatially 'sharp' features (which can be calculated pixel-by-pixel) over complicated and fuzzy features (from pixel clusters);
- Preservation of flexibility in the method by fine hierarchical structuring of the task;
- Supervision and control of preliminary results instead of a single final control, securing a higher total quality and facilitating the revision of the processing method (comp. hierarchical structuring);
- Integration of ancillary data in the classification and/or the validation of (intermediate) results, e.g. by sharp or fuzzy rules;
- Guarantee of a minimum area in every classified segment with complete reference coverage, to avoid undetectable biased results.

From these rules it is becoming clear, that a classification aiming at high quality cannot be performed by a few customised standard functions of a software, but will require additional programming. The following proposals, roughly in a chronological order, are mostly taken from the project 'Land Use Classification of Saxony (FRG) Using Landsat Thematic Mapper Data', which has been completed by 1994 using ERDAS Imagine software as core package (brief descriptions in [30], [31]). The following *aspects of classification* will be discussed:

- General structuring of the classification,
- Tailoring of the image feature space at classification nodes,
- Link to ancillary (non-image) data.

In the beginning, the classification process can, and probably should be *structured into hierarchical levels*, from easy to difficult class separation. The successful completion of a classification at a certain level goes along with a step forward to the next level. The image space decreases. More and more pixels can be excluded from processing (masking).

Easy separation refers to the use of basic image features (local spectral feature vector) with often sufficient information at low levels of the classification tree. However, this statement must be qualified for dependence on the spatial resolution. The higher it gets by progress in sensor design, the less information can be extracted on a pixel-by-pixel level. A direct classification of structurally heterogeneous classes as built-up land is actually based on the abundance of mixels stemming from a large number of different technical surfaces and vegetation. It is exactly a typical signal mix, which allows with some success a functional class assignment based on local image features. But, even a comparatively uniform surface as a grain-field will reveal much more internal variation in an image with the high spatial resolution of the forthcoming sensors, due to unequal growth and interspersed weed: a problem in the use of basic image features.

Difficult class separation will be left for the higher levels of the tree. It refers to the use of more complex image features and feature combinations (like texture, size, etc., comp. above). Sometimes, secondary image features can be treated with standard distance-based classifiers in a common multi-band image. But statistical restrictions have to be observed. Homogeneity criteria for training areas [32] can hardly be fulfilled, when punctual and contextual values form a common data set. Texture or form variables (comp. Figure 3) are rarely normal-distributed. In the contrary, an effective and quick method of structural extraction by filtering aims, as shown in the example below, at a response, which rather reminds to a binary signal (road member or no road member). The fine variations in the bright bands help to find the axis centre. Therefore,

Figure 3. Initial step of road detection by linear filtering, a structural way to classify image data: The filter design is optimised for a road model using form attributes (width, curvature). Left side: input image with approx. 5m resolution; right side: potential motorway candidates [33].

while integrating such techniques, one might alternatively check for different classifiers for the original and the synthetic bands. The individual results can be linked later (e.g. by Boolean expressions).

Broadly one can say, that the proposed classification hierarchy is basically data-driven, always under the premise of a minimum effort and a maximum separability at any node.

With such a general structure being envisaged, one can look at the degrees of freedom at any node of the classification tree (comp. Figure 4):

Referring to the figure, we find the proposed hierarchical structure on the left side of the figure. The original image information (level 0) will be exploited in the levels 1 to 3, and then be recoded or recombined according to the model-driven classification key. All nodes are symbolising individual steps. One basic feature, to be discussed next, is the selection of *feature space contents* at a node. It can best be grouped by its *context level*:

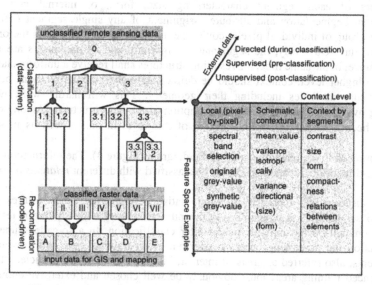

Figure 4. Hierarchical classification structure, selective feature choice and role of external data.

Local features only allow a selection of bands. They might contain original grey or reflectance values, or result from 'vertical' arithmetic operations in the layer stack (e.g. a ratio band). As a further 'local' method, one can transform the original feature vector to a complex of binary spectral shape vectors, a ranking of the response of the original bands instead of the use of absolute values [34]. They might react more robustly in a classification, especially against local low amplitude variations of the reflectance (inhomogeneous surface or imperfection of topographic correction). The 'field of view' of a classifier is limited to a single pixel of a layer stack of selected spectral bands.

The level labelled 'schematic contextual' contains (additional) image information extracted from a moving window of a fixed size. Numerically one can perform a convolution (e.g. a 5x5 high-pass) or a statistical operation (e.g. a 5x5 median). The results can be written to a new layer matching the original matrix size (not regarding problems along the edges). This enlarged, but static 'field of view' might, however, be insufficient to fully explore size-dependent features. Moreover, the 'aperture' of the moving window determines the spatial precision of the extracted information.

The level 'context of segments' is comparatively sophisticated and relies on a preceding image segmentation, which is spatially only limited by the image frame. A segmentation, tested in the present project, had the objective to group pixels into clusters, which are likely representations of real-world objects of what type ever, in other words, without a target-focused specification. The assumption was, that a useful image object (in the sense of land cover classification) can be characterised by low internal contrast, a limited deviation of grey values from a mean value, and clear edges towards the bordering 'objects'. A region-growing approach has been used. For other applications edge-detection or template matching would have been alternative segmentation strategies [35]. Out of the segmentation one will get, first of all, new parameters of each segment characterising size, form, or internal grey value distribution. Segmentation and attribute assignment of any single segment (here still seen as a group of individual pixels) could operationally be applied within the original raster structure. Obviously, the segmentation algorithm and its parameters are highly critical, since, in the same time, the resulting structures shall receive a uniform class tag. Yet, on a higher level, context should be defined through symbolic relations between the segment's attributes including their geometry. This will hardly work without changing over from raster to a vector description of the new structures. Among others, this has been a barrier for the integration of this level in our as well as in similar projects.

We proceed to the remaining axis of the diagram (Figure 4). The information stored in a multi-dimensional feature space can be classified with different *relations to exterior data:*

The unsupervised approach is using statistical clustering, and requires an a-posteriori semantic class assignment to the numeric cluster codes. Exterior reference information will be fed into the model after classification. In a so-called supervised classification the processing order is reversed. From training areas with given class assignments, also referred as areas of interest, image parameters are extracted. The set of referenced training areas must obviously be well chosen and extensive enough, to cover the whole relevant part of the feature space at the present classification step. Moreover, mutual overlaps in the feature space shall be kept as small as possible. Now, within the classification, the numeric class assignment is at the same time a semantic one, since the numeric reference is linked to a training area with known attributes.

With some success, also a combination of the cited methods has been applied; now, the statistical clustering (unsupervised) forms an integrative part of the training area delineation, yielding better grey value distributions and avoiding, to some degree, the disregard of necessary reference (e.g. [36]).

Having discussed an introduction of external data before and after classification, we might have a brief look onto simple strategies, how such data might interact with the classification proper (*directed classification*). Sometimes, we might treat the external information like the image information (as an additional band or layer). Obviously, we must care for common coverage, actuality and spatial resolution and also for a satisfaction of statistical prerequisites of the classifier (e.g. normal-distribution of the reference samples). The approach has frequently been tested in form of a DEM-layer integration [37]; it tries to profit from the fact, that certain land cover classes are centred in elevation zones within a given climatic context. More generally, the objective of a directed classification is a coupling of probability measures from the image and a geographical feature space; the latter is a derivative of prior knowledge about 3-dimensional feature distributions. If a modelling of the distribution pattern is too vague to be introduced, even prior knowledge on the class mixing proportions can improve the classification results by determination of prior probabilities in a maximum likelihood classification [38].

Other related examples are leading over to the post-processing of a (preliminary) classification: One starts with the set-up of a rule base about form, aggregation and association patterns. With its help, 'impossible' or vague class assignments can be detected and corrected. Let us highlight the drainage network as an exemplary target. The classification performance, relying on spectral features only, is limited, for example by width and orientation of the water body, variable reflectance properties (turbid plumes or water plants), or partial screening of the water surface (by tree crowns). Instead of a network of small, linear features we originally get a set of scattered 'water pixels'. Now we introduce a DEM and a rule base about flow direction and the network character of a drainage system. The isolated water pixels and segments might now be linked by an automated tracing of structures in the image and the DEM, that means in a complex information base.

A further example might be taken from the Saxony land cover project: In a mid-summer image (harvest period) small scattered structures aligned along field boundaries are rejected in a classification step for their 'exotic' spectral features. Quite easily, most 'problem pixels' can be addressed as 'mixels', since we had left them unclassified until a late stage of the classification, and find them isolated at the border of large clusters of agricultural land. The high variance of surface characteristics (around 20 basic spectral types for arable land in our case) is correspondingly leading to countless (190) potential class combinations along an edge with, moreover, all different mixing proportions: A classical supervised classification will fail. So, an automated three-step solution has been developed, performing analysis of pixel topology in the classified data, check for a potential spectral mixing situation with signatures from the neighbourhood, and, eventually, assignment to the neighbour class with the highest spectral signature correspondence.

7.5. TRAINING AREAS

The training areas are forming the link between the image feature space and a semantic object in the classification key. Since their definition is most important in a supervised classification, the process shall briefly be discussed. As within the evaluation, one heavily depends on reference information. In a large-area project, a concise referencing of all potential classes and their variations by a field survey is a costly and unlikely case. So external documents will come into the game, but must carefully be evaluated for obvious errors, object definitions different from the classification key (comp. above), and changes, which might have taken place since preparation. Knowledge about geographic patterns and processes is extremely helpful, first to overlook the spectrum of actual land cover, and, second, to assess, if a land use pattern is in accordance with the physical and anthropogeneous disposition of a landscape or not. Examples for the most valuable *reference material* can be given by the following list from the Saxony project:

- CIR aerial photographs 1 : 15 000 (same year as satellite data),
- Spotwise field survey with photographic documentation,
- Official topographic maps 1 : 50 000,
- Punctual field-based crop listings from regional administration of agriculture,
- Selected forest inventory maps from the regional forestry administration.

Apart from available reference, statistical criteria also determine the delineation of training areas: e.g. normal-distribution, standard deviation, spectral distance measures. If a class forms a compact cluster in the feature space, few training samples might do for definition. If disperse, thin 'clouds' prevail, the opposite is true. A data-driven strategy might react through splitting a class into subsets with higher spectral similarity.

Even a careful initial selection of training areas might not prevent the feature space from being insufficiently referenced in some parts. So, operationally, the training area selection will be an *iterative process* (Figure 5), where reference information plays a central role: at start, in localising image segments with known land cover, and, after having extracted the corresponding statistical parameters and having them fed into a classifier, in the verification. When the phase-one results are indicating, that a revision of training areas is commendable, a new statistics set will be compiled from the preceding iteration and the revised set of training areas. A new iteration will be started in an analogue way, until the classification can be accepted.

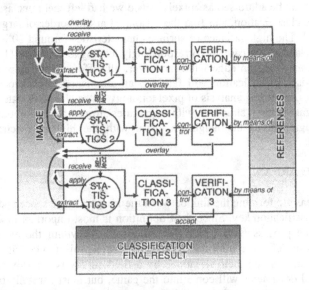

Figure 5. Iterative definition of training areas and training statistics.

7.6. EXAMPLE FOR AN IMPLEMENTATION

Many of the approaches operationally materialised in the processing of the Saxony land use data. The basic structure, though still simplified, has been condensed to the scheme of Figure 6.

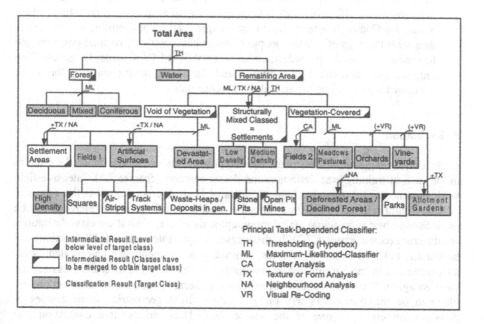

Figure 6. Simplified overview on land use classification scheme and classifiers used for a land use mapping project of Saxony (Germany).

For clearness, the classifiers at the nodes have been aggregated into six groups:

- Thresholding (or hyperbox classifier) and Maximum-Likelihood have been applied on base of training areas within a supervised classification.
- Cluster Analysis as a method of unsupervised classification was basically used to account for the high variation of vegetation and soil signals in the arable land, without aiming at a detection and class assignment of the individual crops (not part of the classification key).
- Texture Analysis was an essential step in the separation of built-up land. A segmentation procedure (comp. chapter 7.4) delivered - among other features - segment size values, which are a quite reliable source to discriminate between signals from post-harvest fields (large segments) and urban structures (small segments). Moreover, allotment gardens with small parcels and a even distribution of buildings and vegetated surfaces could in many cases also be separated from other built-up land with the help of texture measures.
- The term 'Neighbourhood Analysis' was used for post-classification steps treating image clusters with unclear assignment. Examples are the assistance in declined

forest classification (forest understorey signal is similar to barren farm land) by a relative situation next to closed forest stands, the regional context (indicating a high percentage of forest damage in forestry documents), the size and the appearance in topographic data. Similarly, the neighbourhood context could be used in the problem case built-up land to eliminate single pixels miss-classified as fields or meadows. If additionally an integration of a digital road-network is possible, a very effective check is looking for a road access to any built-up cluster.

- Visual Re-Coding is a term, which shall express all manual editing applied to the data with the help of existing maps. In some cases with a low total coverage, no formalised automatic procedure could be established (Vineyards, Orchards). The only escape was a visual interpretation and class assignment comparing the image structures with topographic and other reference data.

8. Evaluation

Evaluation for large areas is restricted by man power. The situation is much the same as in the case of training area definition and documentation (chapter 7.5). Intensive field work cannot cover thousands of square miles of land. A crucial point is the access to reliable digital reference data. In a fortunate case, a historic land use reference can be used. Second best is an actual digital topographic data base, even if the class definitions are different (comp. chapter 7.2). In these cases a layer intersection (GIS function) can be started, and, in a follow-on process, conflicting assignments can be revealed. Now, all concentration can be directed to the conflicts, and only the affected areas can be checked again. It is obvious, that some change patterns, generated by intersection, are likely to be realistic, others will rather indicate slight geometric inconsistencies or classification errors in one of the source data. Thus, an iterative evaluation and classification could be started, aiming at the solution of the conflicts. For the quantitative aspect of quality assessment, a good overview is given by [39].

In case of an initial classification of a large area and analogue references, an evaluation:

- will account only for evident errors in the total classified area (requiring expert knowledge about the landscape and its characteristic patterns)
- will always concentrate on small statistical samples with extensive reference information for a numeric assessment of accuracy measures (which cannot be derived by means of the correct assignments of the training areas used for the supervised classification)
- can make plausibility tests using 'weak' reference information (e.g. DEM, medium scale plant production suitability map for agriculture, tree species distribution data for the forestry, etc.).

A detailed evaluation could be performed in the case of a land cover change study for the comparatively small Dresden agglomeration [40]. The article also gives figures of land use change rates in time for this region, what must be connected to actuality discussions [29].

The situation in many European areas has been improved during the last couple of years, since digital land use data have been operationally produced (especially CORINE Land Use [41]); even if they are not fully exploiting the spatial resolution capabilities of the image data, their suitability as a reference source would be undisputed.

9. Map Generation

The completion of the classification will not be the end of the project, if the results shall be fit for use. In the initial form the classification is a thematic raster containing class codes and header information about the orientation, the geodetic projection and the resolution of the data. Minimum requirements for a digital distribution must be the export into a variety of raster formats as well as a provision of *meta-data* containing:

- the scope of the project,
- the source data, including most important the date of data take (= temporal reference),
- the classification key with precise specifications,
- a decoding table to transfer a numeric code to verbal class definitions,
- basic remarks on the classification methods and the accuracy achieved,
- geometric and geodetic references including scale and accuracy assessment,
- producer, distributor and copyrights.

However, as a useful supplement to the digital data base, the production of a thematic map series should be envisaged. The statement 'planning is still done on the base of maps' [42, p. 66] is 20 years later as actual as it was. The maps should obviously show the mentioned meta-data along with the classification, but must also contain topographic elements for a localisation and thematic relation of the theme. This can nowadays be achieved with low manual interaction, when topographic information is at hand in a digital form. Large-scale *topographic vector data bases*, also called 'digital landscape models', with a detailed feature separation are available for many European countries, e.g.:

- ATKIS for the German territory by the stately surveying agencies [43],
- MERIDIAN for the UK by the British Ordnance Survey [44],
- BD TOPO for France by the Institut Géographique National [45],

Now, the focus must be directed to the actual map design, which implies the following questions:

- appropriate map scale and sheet division,
- colour scheme for the thematic classes,
- selection of topographic features and associated signatures,
- contents and layout of the map frame.

When the map face shall show the ungeneralised classification, the reproduction scale has to allow a pixel-sharp identification on the one hand, but a 'stepped' appearance of class contours (by the raster structure) has to be avoided on the other hand. A 1 : 100 000 scale (corresponding to a pixel size of 0.25 x 0.25 mm^2) has proved to be a good compromise. This is in good accordance with recommendations by [46]. Guidelines for the general layout of the 'Saxony example' (including sheet size, coverage, composition of map face and frame elements) have been derived from the official topographic maps of that scale. Its *colour scheme* for the thematic classes is reflecting principles like:

- auto-plausibility (green forest, blue water),
- principal groups of classes clumped to clusters in the colour space,
- natural colours used for classes of dominantly natural land cover, and colours, rare in nature, for dominantly technical land cover (built-up land, communication lines),
- low-saturated colours assigned to compact, widespread classes, high-saturated colours to small, scattered classes.

Only a reduced drainage network, major roads and rail-links, and some lettering have been extracted from the topographic data; the thin, linear elements do not cause much visual conflict with the classification. For a sufficient contrast, all additions are printed over in black. But, above what has been realised in our example, one should additionally account for the strong correlation between relief and land use in a map.

For very pragmatic reasons, the well-established general German topographic map layout has been adapted. The title page just shows the individual guide colour yellow (1:100 000 topographic map: red). All map frame contents have been grouped into static parts (identical in all sheets) and sheet-specific parts to minimise production effort. Meta-data of the classification (comp. above) are integrated in the map frame, as well as a brief introduction in the method of digital land use classification. An additional feature is a pie chart of the class frequencies for each individual sheet. An idea of the map design will be given by Figure 7.

10. Land Use Monitoring in Less-developed Countries (Albania)

Land use classification with optical remote sensing data as an actual, relatively unbiased, low-cost method is widely accepted and operational in 'first-world countries'. A fully digital satellite image processing must be seen as state-of-the-art. Even if the results will never be error-free, an internal consistency can be achieved much easier, compared to a team work on the same task using analogue data.

But why should less-developed countries invest time and money in image-processing (and GIS-) techniques? Only very general answers can be given, which (hopefully) should also be valid for the Albanian situation: We assume, that, generally, land use information is sensible data, even more so, if there is a high pressure by environmental problems:

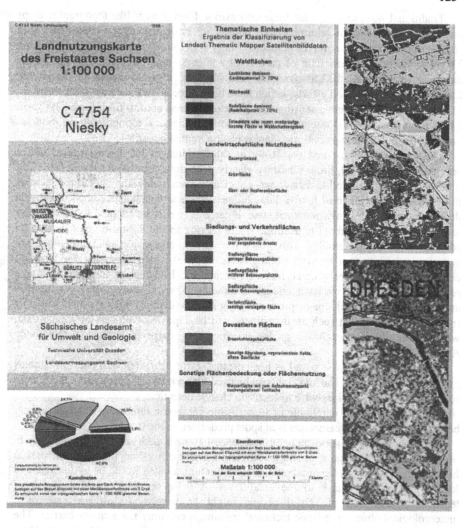

Figure 7. Design of Series '1 : 100 000 Land Use Map of the State of Saxony'. Upper left: title page, lower left: pie chart of class frequencies, upper middle: thematic legend, lower middle: coordinates and scale, upper right: section of map face from sheet 'Niesky' in reduced scale, lower right: section of sheet 'Dresden' in original scale. Please see appendix for image in colour.

A 1 : 100 000 scale land use data set is detailed enough to picture patterns and problems in a regional scale. This is more detail and a higher resolution than what we have at hand with most other environmental data (e.g. climatic data, air and water-pollutants, soil characteristics, geology, ...) nearly all over the world. Where problems culminate, the data can, at least, spatially guide a large-scale, parcel-by-parcel survey with an implication on any individual proprietor. Until now, the latter still requires aerial photographs and/or ground surveys. But with the upcoming satellite systems, we might soon be able to handle some information problems of the local scale from space, too.

Traditional ways to access land-use information from sources like topographic maps and statistics by various administrative bodies are critical; care for consistency and regular updates is often severely neglected under poor economic conditions. Following Talani [47], the Albanian state of cartography allows some optimism because of the existence of an established topographic base covering all scale groups in the proved former Soviet system, and efforts, which have been directed to their transfer it in an up-to-date digital form. Remote sensing data has already been used to fill information gaps. There is no reason to suppose a competition between classical topographic and remote-sensing-based data. In the contrary, there is clear a mutual benefit.

The general low cost and versatility of satellite images has been stressed in the article of Konecny [48], the availability clearly been demonstrated by Baetz [49]. The multi-purpose of the satellite data comes out clearly even by looking only at the lectures in this issue. An archive of image data, augmented on a regular base, is an invaluable dated document of an environmental state. It can even easily be adapted to find another useful application in school education. While staying in the digital domain from data acquisition to digital results, also costs for copies, processing, and archiving are comparatively low (not so with photographic material, since geometry and brilliant colour reproduction are crucial for an interpreter). Looking at the digital image processing equipment, we have enough computing power already with low-cost PCs and only a professional image processing package will imply major costs. But it is important to know, that such an investment will be capable of processing image data, which more or less excludes analogue processing (hyperspectral data, Synthetic Aperture RADAR).

Once, a digital land use base has been created, efforts directed into later-on updates can be greatly reduced. The 'old' data can be used for an image segmentation, the feature space can be analysed class-wise to focus on change detection.

The old problem of unreliable or inaccessible geometric references can quite easily be overcome by GPS measurements at spots predefined in the images.

The importance of good thematic reference has been addressed. When looking in particular at phenological data and a sufficient number of well-documented test sites, a highly important assistance of a land use classification, an idea would be to delegate this long-term monitoring task to the local level. A volunteer force of students, local teachers or any sort of skilful people might be trained. As in the case of a meteorological observer, a basic technical training is certainly not a major barrier. The crucial point is the reliability and devotion to take a share in environmental monitoring.

If a little plus in environmental awareness and technical skills in the field of geo-data processing could be spread within a land-use project, all effort will obviously be justified.

11. References

1. Gierloff-Emden, H.-G.(1989): *Fernerkundungskartographie mit Satellitenaufnahmen.- Allgemeine Grundlagen und Anwendungen*. Bd. IV/1 of 'Enzyklopädie Die Kartographie und ihre Randgebiete', Franz Deuticke, Wien.
2. Prechtel, N. (1996): *Flächennutzungskartierung mit Satellitenaufnahmen*. Wissenschaftliche Zeitschrift der TU Dresden, 1/96, pp. 62-66.

3. Wieneke, F. (1988): *Satellitenbildauswertung - Methodische Grundlagen und ausgewählte Beispiele.* Münchener Geographische Abhandlungen, **A38**, 169 p.

4. Skole, D. (1996): *Land Use and Land Cover Change.* - The Earth Observer, 8/3, EOS, pp. 36-40.

5. Konecny, G. (1995): *Satelliten-Fernerkundung und Kartographie.* - Geo-Informations-Systeme, 2/1995, pp. 3-12.

6. Bodechtel, J. and Zilger, J. (1995): *MOMS - History, Concepts, Goals.* - In: Lanzl, F. (ed.) Proceedings of the MOMS-02 Symposium, Cologne, July, 5th-7th, 1995, EARSeL, Paris, pp. 12-25.

7. Kontoes, C. and Stackenborg, J. (1990): *Availability of Cloud-Free Landsat Images for Operational Projects.* International Journal of Remote Sensing, **11/9**, pp. 1599-1608.

8. Orthaber, H. (1999): *Bilddatenorientierte atmosphärische Korrektur und Auswertung von Satellitenbildern vegetationsdominierter Gebiete.* Kartographische Bausteine **16**, Dresden, in print.

9. Lieth, H. (1974): *Phenology and Seasonality Modelling.* Ecological Studies, 8, Springer, Heidelberg/New York, 444 p.

10. Deutscher Wetterdienst (DWD). *Agrarmeteorologischer Wochenhinweis für das Gebiet der Bundesrepublik Deutschland,* Offenbach. Comp. http://www.dwd.de/services/lw_home.html. Phänologische Beobachtungen, comp.: http://www.dwd.de/services/LW1/WOHIWEI/muster/intrawo.html.

11. Price, J. C. (1994): *How Unique are Spectral Signatures?.* Remote Sensing of Environment, **49**, pp. 181-186.

12. Steiner, D. (1961): *Die Jahreszeit als Faktor bei der Landnutzungsinterpretation.* Landeskundliche Luftbildauswertung im mitteleuropäischen Raum, **5**, Bundesanstalt für Landeskunde und Raumforschung, Bad Godesberg.

13. Preissler, H., Bohbot, H., Mehl, W. and Sommer, S. (1998): *MedSpec - A Spectral Database as a Tool to Support the Use of Imaging Spectroscopy Data for Environmental Monitoring.* In: Schaepman, M. Schläpfer, D. and Itten, K. eds.): 1st EARSeL Workshop on Imaging Spectroscopy. EARSeL, Paris, pp. 455-462.

14. Price, J.C. (1995): *Examples of High resolution Visible and Near Infrared Reflectance Spectra and a Standardized Collection for Remote Sensing Studies.* Intern. Journal of Remote Sensing, **6/95**, pp. 993-1000.

15. http://www.geog.ucl.ac.uk/~mdisney/brdf.html (BRDF-related sites on the WWW).

16. Meier, U. (1999) *Growth Stages for Mono- and Dycotyledonous Plants.* BBCH-Monography Federal Biological Research Centre for Agriculture and Forestry, Parey, Hamburg / Berlin.

17. http://earth.esa.int80/aboutdescw.

18. Costa, N. (2000): *WWW Information Services for Earth Observation and Environmental Information.* This issue.

19. Ehrhardt (1990): *Modellorientierte Entzerrung von Thematic-Mapper-Rohdaten.* DLR Fachbericht **90-55.** Oberpfaffenhofen, 103 p.

20. Lei, Q., Henkel, J. Frei, M., Mehl, H. Lörcher, G. and Bodechtel, J. (1995): *Radiometric Noise Correction of Panchromatic High Resolution Data of MOMS-02.* - In: Lanzl, F. (ed.) Proceedings of the MOMS-02 Symposium, Cologne, July, 5th-7th, 1995, EARSeL, Paris, pp. 303-307.

21. Buchroithner, M.F. (1989): *Fernerkundungskartographie mit Satellitenaufnahmen.- Digitale Methoden, Reliefkartierung, geowissenschaftliche Anwendungsgebiete.* Bd IV/2 of 'Enzyklopädie Die Kartographie und ihre Randgebiete', Franz Deuticke, Wien.

22. Banon, G.J.F., Barrerea, J. (1989): *Morphological Filtering for Stripping Correction of SPOT Images.* Photogrammetria, **43**, pp. 195-205.

23. Hill, J. (2000): *Resource Assessments and Land Degradation Monitoring with Earth Observation Satellites.* This issue.

24. Richter, R. (1996): *A Spatially-Adaptive Fast Atmospheric Correction Algorithm.* ERDAS IMAGINE - ATCOR2 User Manual. Geosystems (Distr.), Germering.

25. Toutin, T. (2000): *Map Making Using remote Sensing Image Data.* This issue.

26. Almer, A., Raggam, J. and Strobl, D. (1991): High-Precision Geocoding of Spaceborne Remote Sensing Data of High-Relief terrain. Proceedings ACMS/ASPRS/AutoCarto Annual Convention, Baltimore, Ma, March 25th-29th.

27. Todd, W.J., Gehring, D.G., Hamann, J.F. (1980): *Landsat Wildland Mapping Accuracy.* Photogrammetric Engineering and Remote Sensing, **46/4**, pp. 509-520.

28. Anderson, J. R., Hardy, E. E., Roach, J. T. and Wittmer, R. E. (1976): *A Land Use and Land Cover Classification System for use with remote sensing Data.* USGS Professional Paper, **964**, Washington.

126

29. Prechtel, N. (1997): *CORINE-Bodenbedeckungsdaten für Ostdeutschland aus Anwendersicht*. Zeitschrift für Photogrammetrie und Fernerkundung, **3**, pp. 92-101.

30. Prechtel, N. (1995): *A Comprehensive 1 : 100000 Land Use Map Series for the Federal State of Saxony (Germany) from Satellite Imagery*. - Proceedings of 17th International Cartographic Conference in Barcelona, Sept., 3rd-9th, Vol. II, ICA, pp. 2281-2287.

31. Prechtel, N. (1996): *Flächennutzungskartierung mit Satellitenaufnahmen*. Wissenschaftliche Zeitschrift der TU Dresden, **45/1**, pp. 62-66.

32. Schulz, B.-S.(1990): *Analyse der statistischen Voraussetzungen zur Klassifizierung multispektraler Daten*. - Zeitschrift für Photogrammetrie und Fernerkundung, **3/1990**, pp. 66-74.

33. Prechtel, N. und O. Bringmann (1998): *Near-Real-Time Road Extraction from Satellite Images Using Vector Reference Data*. International Archives of Photogrammetry and Remote Sensing, Vol. 23, Part 2 = Proceedings of the Commission II Symposium Data Integration: Systems and Techniques, Cambridge, July 13th – 17th, pp. 229-234.

34. Carlotto, M.J. (1998): *Spectral Shape Classification of Landsat Thematic Mapper Imagery*. Photogrammetric Engineering and Remote Sensing, **64/9**, pp. 905-913.

35. Steinbrecher, R. (1993): *Bildverarbeitung in der Praxis*. Oldenbourg Verlag, München.

36. Gangkofner, U. (1996): *Methodische Untersuchungen zur Vor- und Nachbereitung der Maximum Likelihood Klassifikation optischer Fernerkundungsdaten*. Münchener Geographische Abhandlungen, **B27**.

37. Rössler, G. (1989): *Almvegetationsklassifizierung mit Satellitenbildern*. Diploma Thesis, Vienna University of Technology, Chair of Photogrammetry and Remote Sensing.

38. Gorte, B. G. H. (1999): *Local Statistics in Supervised Classification*. In: Nieuwenhuis, G. J. A., Vaughan, R. A., Molenaar, M. (eds.): Operational Remote Sensing for Sustainable Development. Balkema, Rotterdam, pp. 325-330.

39. Congalton, R. G. (1991): *A Review of Assessing the Accuracy of Classifications of Remotely Sensed Data*. Remote Sensing of Environment, **37/91**, pp. 35-46.

40. Meinel, G., Knapp, C., Buchroithner, M. und N. Prechtel (1995): *Detection of Ecologically Relevant Landuse Changes in the Agglomeration of Dresden Using Multitemporal Landsat-TM-Data*. - In: Proceedings of EARSeL Workshop on Pollution Monitoring and GIS in Brandys nad Labem, May, 15th-18th, pp. 158-167, EARSeL, Paris.

41. Büttner, G. (2000): *Land Cover - Land Use Mapping within the European CORINE Programme*. This issue.

42. Lüttig, G. (1979): *Geoscientific Maps as a Basis for Land Use Planning*. Geologiska Föreningens i Stockholm Förhandlingar, **1**, pp. 65-69.

43. Arbeitsgemeinschaft der Vermessungsverwaltungen der Länder der Bundesrepublik Deutschland (AdV) (1988): *ATKIS-Gesamtdokumentation*, Hannover.

44. http://www.ordsvy.gov.uk/home/index.html: - Homepage *British Ordnance Survey*, Southampton.

45. http://www.ign.fr/ - Homepage *Institut Géographique National*, Paris.

46. Albertz, J. (1994): *Beiträge der Satelliten-Fernerkundung zur topographischen und thematischen Kartierung*. - Bollmann et al. (eds.): Umweltinformation und Karte. Sonderband zum 43. Deutschen Kartographentag in Trier. pp. 27-35.

47. Talani, R. (1998): *Kartographie in Albanien*. Kartographische Nachrichten, **6/98**, pp. 228-234.

48. Baetz, W. (2000): *Current Remote Sensing Availability*. This issue.

49. Konecny, G. (2000): *Overview of Possibilities with Future Satellite Image Data*. This issue.

EXPERIENCES WITH THE IMPLEMENTATION OF GIS AND REMOTE SENSING IN THE CZECH FOREST MANAGEMENT

TOMAS BENEŠ
ÚHÚL Forest Management Institute,
Nábrezní 1326,250 44 Brandýs nad Labem,
Czech Republic
E-mail: benes@uhul.cz

Abstract

Currently in all former "socialist countries" a technological revolution is going on: information technology development and the implementation of new digital sources of data utilisation. In all European countries forestry is a major part of this market.

But especially in the above mentioned countries which are in transition, a renovation of the forest management is necessary, under the scope of the new laws about private forest property. Geographic Information Systems (GIS) have been discovered as new and fast tools for modern forest management.

The Czech forestry, faced with typical central European environmental problems due to air pollution and forest diseases, became a leader in remote sensing and GIS technological applications in the Czech Republic. The history of "GIS forestry" is relatively long, beginning in 1988. The current situation of GIS implementation in Czech forestry, the traditional and new data sources, and future perspectives in the context of the state administration are illustrated in this paper. Emphases will be put on modelling, monitoring and forecasting of forest diseases using a technology based on the Landsat (TM) image classification and detailed-data GIS data base generation.

1. Background

Like in other Central Eastern European countries in transition, previously the Czech forestry was a centrally planned branch of the national economy. Following the big changes in 1989, new political rules took place. The most important one, not for forestry only, was the establishment of private property. The creation of new laws and their governmental and parliamental confirmation is a very difficult and long process. In 1997 the new forest law was established as a one of the concrete results of this process. This implies the implementation of many new facts and relations into the "old forest management strategy". In a long discussion, GIS technology was considered the most useful tool to realise all changes and information needs concerning the new forest management.

M.F. Buchroithner (ed.),
Remote Sensing for Environmental Data in Albania: A Stragegy for Integrated Management, 127–132.
© 2000 *Kluwer Academic Publishers. Printed in the Netherlands.*

2. Technology

The new technology line for forest management can be divided into the following segments:

1. data sources
2. database generation
3. data management
4. data processing
5. output for practical work.

2.1. DATA SOURCES

Besides traditional data sources for the forest management process recent aerial and satellite images are used: a concrete example of remote sensing data application in practice. Which types of remote sensing data are used? First of all, mono- and stereoscopic aerial metric photographs, because they yield a high degree of detail. Furthermore, satellite data from Landsat TM are also used. They are especially applied to forest diseases monitoring. Both types of these objective and up-to-date data sources are used in digital form only, which are, in general, bulk-processed. For forest diseases satellite images are processed by a special methodology.

The second major data source is represented by the documentation of the forest management plan. They consist of two main parts: forest maps and forest inventory books. The thematic forest maps comprehend topological forest maps, typological-, forest inventory-, forest stand maps, etc. The basic scale of these maps is 1: 10 000. They have all, archive material included, been transferred into digital form.

As a numerical data source the forest inventory book has been organised using the Fox plus program, but is currently rebuilt under the Oracle database system.

2.2. DATABASE GENERATION

Database generation is currently realised on the basis of the Oracle software, where clearly connected graphical and numerical data exist. To illustrate the very rich content of this database it shall be mentioned, that the Czech database sequence has up to 53 items for each forest stand, which comprehends approximately 20 hectares in average. The necessary parameter selection is realised by keywords.

2.3. DATA MANAGEMENT

The data management of our system is materialised by the Czech GIS software system package TopoL. This system has been generated for both large-size data management, processing, and thematic output creation including printing. The structure, philosophy and main outputs of TopoL would require a separate discussion. To briefly illustrate its parameters only the special modules for the export/import to/from Arc/Info and Intergraph shall be mentioned. These are very important tools for the communication between TopoL and the world-wide most used GIS systems.

In addition, a special version for photogrammetric and thematic airphoto evaluation shall be mentioned. They are called PhotoL, Special Data Viewer TopoL and Topotax, a special program for forest inventory in the field, etc. Generally speaking, the full version of the TopoL software package covers not only the GIS and remote sensing and various other data applications, but especially highly detailed forest modelling.

2.4. DATA PROCESSING

The full-function technology line has been created by the Czech public forest managers. Subsequently this line has been adapted for practical needs of private owners. The Information Data Centre (IDC) of UHUL, the Czech Forest Management Institute, was recently established as a supervisor for the computer technology application in forestry. Two parallel lines were developed: first for private forest owners, who own more than 50 hectares of forestry, because the new forest law requires a forest management plan for this category of forest owners. The second line has been generated for supervision and up-dating of forest databases and fast selection of data. This version supports current state targets or other requirements. Both lines are illustrated in Figure 1.

2.5. OUTPUT FOR PRACTICAL WORK

The main output of this technology system are thematic maps displaying the current state of forest lands in the Czech Republic. These maps, from their general appearance, legend, colours and content the same as the old ones, have significant differences to those. Their generation technology is very much different, because they are completely realised by digital processing.

The main advantages are:
- fast reaction to the current needs of the customers
- more operationalisation of all processes of map generation in comparison to the traditional way
- very low level of costs
- map sheets are printed in the required quality
- real-time generation of required thematic maps.

Due to the continuous privatisation in the Czech Republic, currently the forest management plans are mostly created by authorised private firms. In spite of this, the UHUL Institute has presently a lot of work to do, and this will not change in the future. Why? A new form of state forest management was created by the Ministry of Agriculture. This is the Regional Plan for Forest Development (RPFD), which was established in response to the expected membership of the Czech Republic in the EU. The main differences between the old and the new Forest Management Plan are the following:
- Data sources for optimum objectivity, i.e. remote sensing data, play the most important role.
- All aspects which are necessary for sustainable forest development are included.
- The state of the forests is currently documented in maps.
- Continuously a new type of thematic forest map, the typological map is created.

The latter one is processed for new regions, i.e. forest nature districts in spite of FMPs.

This practically new level in the Czech forest management documentation displays a new dimension of quality – the balanced weight between the production function and other functions of the forest.

As an example, that these new RPFDs are the most objective form for forest documentation, modelling and forecasting can be used. It is realised by a unique combined technology for data observation, investigation and acquisition. All these data are processed by a generic GIS technology for all parts of the Czech territory. Their presentations, including printing, are created by a unique printing technology. This special printing method was developed in co-operation of UHUL and the private firm TopoL-Pro Ltd. When the digital technology and GIS were initialised at UHUL, the special branch of the Information and Data Centre (IDC) was established. Its structure and tasks are illustrated in Figure 2.

Other outputs, belonging to the forest inventory and forest statistics were up-dated by the new technology concerning both, the classical information, but also other new data, e.g. forest diseases derived from satellite data, typological characteristics etc.

3. Conclusion

The described technology of forest management is of high importance not only for the Czech Republic, but for all former "socialist countries" in transition. Thus, it has strong effects on forestry in general. Consequently, the new technology was officially established by the Czech Ministry of Agriculture. The outreaching of the developed principles into other branches of land management has already started. The presented approach is very cost-effective, operational and does not require a lot of money. Currently, a general invasion of GIS technology into all parts of the land market in the Czech Republic can be observed.

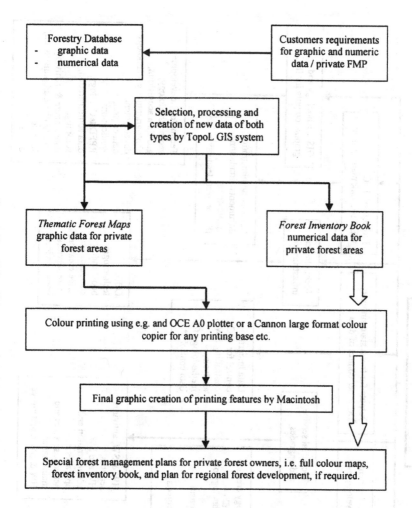

Figure 1. Technology line for private owners

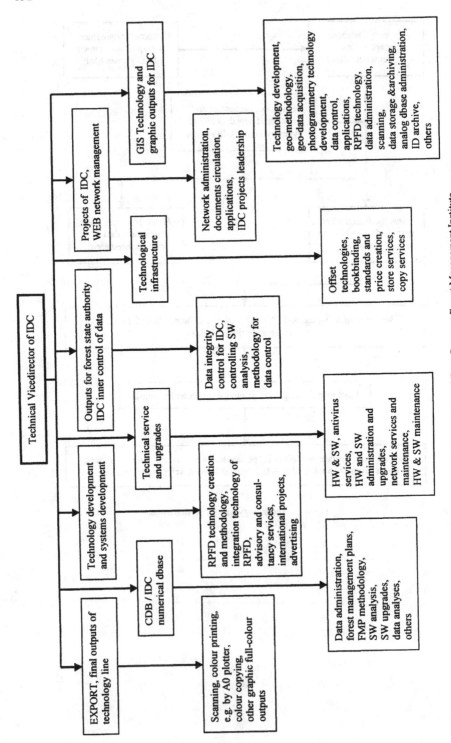

Figure 2. Flow chart of Information and Data Center at Forest Management Institute.

RESOURCE ASSESSMENTS AND LAND DEGRADATION MONITORING WITH EARTH OBSERVATION SATELLITES

JOACHIM HILL
Remote Sensing Department
University of Trier
D-54286 Trier, Germany

Introduction

Land degradation processes which imply a reduction of the potential productivity of the land (e.g., soil degradation and accelerated erosion, reduction of the quantity and diversity of natural vegetation) are widely spread in the Mediterranean basin. As a continuation of the long history of human pressure upon land resources, the main environmental impact originates from interactions between climatic characteristics and ecologically unbalanced human interventions. which, in the sense of recent definitions of the United Nations Environmental Programme[1], are often summarised as *desertification processes* (e.g., Thomas and Middleton, 1994). An overview of the ecological, physical, social, economic and cultural issues which are collectively contributing to the increasing risk of further degradation of Mediterranean lands has been presented by Perez-Trejo (1994). The same author concludes that a reconceptualisation of desertification - one more appropriate for the European situation - is needed in which the role of urban-industrial expansion, tourism and agriculture in relation to the allocation of water resources are seen as significant contributors to the problem. Inadequate land use practises (e.g., excessive grazing, fuelwood collection, uncontrolled fires) further contribute to the acceleration of degradation processes which result primarily from complex interactions of plant growth and erosion processes. It is now widely agreed that accelerated water erosion is one of the most important sources of soil degradation; an average yearly soil loss of more than 15 tons/ha was reported by Grenon and Batisse (1989) to occur in more than one third of the Mediterranean basin. This excessive loss of soil, nutrients and seeds from the ecosystem hampers also the regeneration capacity of the vegetation by disrupting plant-water relations, and thus drives a mechanism of sometimes irreversible environmental damage. If erosion is not

---

[1] UNEP (1991) has redefined desertification in the sense of considering it as "the land degradation in arid, semiarid and dry-semihumid areas resulting mainly from adverse human impact" where the term 'land' includes land and local water resources, the land surface and its natural vegetation.

133

M.F. Buchroithner (ed.),
Remote Sensing for Environmental Data in Albania: A Stragey for Integrated Management, 133–152.
© 2000 *Kluwer Academic Publishers. Printed in the Netherlands.*

stopped, further increase in runoff, sheet and gully erosion on sloping ground may ultimately destroy the productive value of the land.

1. Resource Inventories and Degradation Processes

Inventories of available resources, but also the identification and continuous monitoring of degradational processes form the basis for drafting and implementing rational development plans for a sustained use of land resources, such as it is needed for countries like Albania in particular. However, the precise localization of initial vegetation damage and erosion processes requires a spatial resolution that is adequate for an integration with existing topographic maps at local to regional level (i.e. scales 1:50 000 to 1:200 000). Conventional mapping approaches (i.e. field surveys, air photo interpretation), although important and not replacable for local studies, can not provide these detailed mapping results for larger areas. This is mainly due to prohibitive costs, deficient mapping quality in difficult or inaccessible terrain and insufficient standardization and repeatability of these surveys.

The advantages of satellite remote sensing result from its synoptic nature, comprehensive spatial information and objective, repetitive coverage. While it has initially been primarily used for resource mapping and inventory, it turns out that retrospective analysis, monitoring and predictive modelling is becoming increasingly important. Because of the availability of long data records and methodological advances, remote sensing with earth observation satellites has already become an irreplaceable component of resource assessments and monitoring strategies for observing and quantifying environmental change. Landsat data in particular provide a continuous, consistent quality record of continental surfaces, dating from 1972.

2. Instrumental Opportunities and Methodological Issues

When compared to the early 1970s when the first Landsat system was placed into a space orbit remote sensing systems meanwhile exhibit an enormous diversity in terms of spectral, spatial and temporal parameters (Kramer, 1996), and it depends on the user requirements which system is to be used. If a frequent repetitive coverage with relatively low spatial resolution is desired (e.g., for meteorology) one would certainly be inclined to base the approach on the AVHRR-system available from the polar-orbiting satellites of the NOAA series, or even on data from geostationary satellites such as METEOSAT or GOES. Alternatively, one might look for the highest spatial and spectral resolution available, even at the expense of relatively low repetition rates, and would thus choose one of the available earth observation satellite systems (Landsat, SPOT, IRS, IKONOS). With regard to the introductory papers in this volume we shall not further discuss this issue.

However, although it is agreed that remote sensing provides a convenient source of information, the problem is that the data collected by these instruments do not directly correspond to the information we need. We must therefore interpret the signal which has interacted with remote objects in terms of the properties of these remote objects (Verstraete, 1994). Depending on the scientific perspectives, quite some variety of approaches and strategies has been proposed for assessing environmental characteristics of arid, semi-arid and sub-humid ecosystems based on remote sensing (see, for example, Hill and Peter, 1996). The different perspectives expressed therein are not only emphasising methodological preferences but, more fundamentally, also reflect important paradigms of scientific disciplines.

2.1. REMOTELY SENSED PRIMARY PARAMETERS; THEMATIC CONCEPTS AND DERIVED INDICES

Before discussing appropriate scene models one needs to understand that the initial part of the processing chain (i.e., data pre-processing) deals with the geometric rectification of digital imagery and with turning uncalibrated image grey values into physical quantities. Engineering data about the detector sensitivity (i.e., calibration coefficients) permit to reconvert encoded DN into measured radiance, and radiative transfer calculations can be used to correct for atmospheric effects, such that the surface-reflected radiance is restored from the satellite-measured signal. Dividing this term by the downwelling solar irradiance provides us with an important primary parameter which is termed bi-directional reflectance (Figure 1); albedo and surface temperature are other primary parameters which, as a result of similar processing chains, can be derived from optical remote sensing systems. Here, not much emphasis will be given to the radiometric processing of satellite images nor to geometric rectification issues because both involve routine operations which are either described elsewhere (e.g., Hill, 1993a; Hill et al., 1995) or in other contributions to this volume.

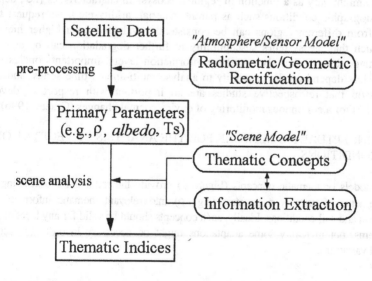

Figure 1. The conversion of satellite raw data into standardised thematic information layers.

In order to assess land resources or land degradation processes it is necessary to define diagnostic indicators. These may be primary indicators (e.g, high salt content in salinised soils) or secondary indicators which are produced by the problem (e.g., reduced vigour of vegetation). The most useful indicators are those which have distinct spectral signatures and are unique to a particular issue (Johnston and Barson, 1990). Although of relevance to the global radiation budget, albedo and reflectance changes per se are not direct indicators of land degradation processes, in particular when we consider spatially complex areas like the European Mediterranean. It should be recalled that, under Mediterranean conditions, changes in albedo (surely of importance as indicator for the desert encroachment in the African Sahel or similar regions) might be due to changes in land surface characteristics which do not necessarily imply negative effects (i.e., increase of greenhouse areas, maturing cereals, non-photosynthetic vegetation, clearing of woodlands, etc.). It is therefore not sufficient to simply map albedo changes over time, but we need to infer the environmental impact of altered reflectance properties by characterising their physical nature in terms of land surface conditions. We thus need an appropriate scene model which can be used to convert multi-spectral reflectance into thematic information (Figure 1). A variety of methods have been proposed which range from empirical spectral indices to the design and inversion of physically-based models. While the applicability of the various approaches depends on the nature and accuracy of the desired information and the availability of resources (i.e., sensor characteristics), an important prerequisite for their operational use is that they must also satisfy specific requirements in terms of standardisation and portability.

Both the development of suitable indices and their interpretation in the thematic context of land degradation monitoring requires a conceptual framework which allows to draw concise conclusions about the land surface conditions. Though these underlying concepts might vary as a function of regional ecosystem characteristics (i.e., depending on physiographic conditions such as parent material, aridity etc.), we request that the results from different regions can be consistently evaluated on a higher hierarchical level, such that the system's susceptibility to further degradation can be assessed by using image-derived, but also ancillary information layers. Important conclusions will nevertheless depend on the capability to analyse multi-annual time series, and it is for this reason that retrospective studies are so important with respect to developing approaches for a continuous monitoring of environmental changes (Graetz, 1996).

2.2. CONCEPTUAL FRAMEWORK FOR THE ANALYSIS OF VEGETATION AND SOIL CONDITIONS

Scene models or thematic concepts (Figure 1) provide the rational for translating remote measurements of primary parameters (e.g., ρ) into relevant thematic information about vegetation and soil conditions. Ideally, such concepts should be valid for any location in large ecosystems, but in reality some adaptations might be necessary in order to account for regional variations.

2.2.1. *Mapping Soil Conditions.*

Eroded soils are often recognised through typical soil colour changes which are due to the removed topsoil. It is nevertheless difficult to define universally applicable concepts which account for a variability of soil types and the corresponding sequence of pedogenetic horizons. An important approach refers to basic concepts which consider soil development to be either *progressive* or *regressive* with time (Birkeland, 1990). Under progressive development, soils become better differentiated by horizons, and horizon contrasts become stronger. In contrast, regressive pedogenesis refers to the addition of material to the surface at a rate that suppresses soil formation (i.e. eolian dunes, glacial moraines, distal fans, etc.), or the suppression of pedogenesis and the truncation of soil horizons by surface erosion (Figure 2). Both, progressive and regressive pedogenesis cause alterations of the soil surface which, due to corresponding colour changes, are detectable through the wavelength-dependant variations of ρ (e.g., Baumgardner et al., 1985; Escadafal, 1993). The intensity of brunification and rubification, and the organic matter content of the topsoil material thus provide important diagnostic features for the spectral identification of a majority of undisturbed Mediterranean soils (e.g., cambisols, fluvisols, luvisols, vertisols, rendzinas).

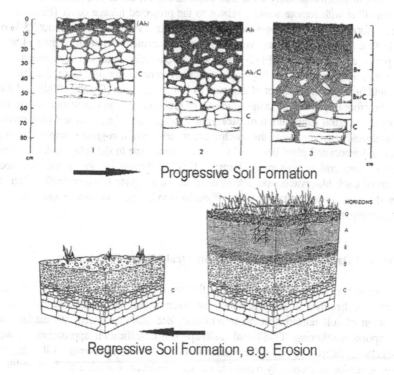

Figure 2. The concept of progressive (top) and regressive soil formation (bottom) as outlined by Birkeland (1990)

Compared to that, soil erosion produces truncated soil profiles which are characterised by decreasing amounts of iron oxides and organic carbon, while the proportion of parent material increases (e.g., lithosols, regosols). Most parent materials differ spectrally from developed soil substrates, in particular due to specific spectral absorption features and increased albedo levels.

The resulting concept, which is based on the spectral contrast between developed substrates and parent materials, seems to provide a widely applicable framework for relating spectrally detectable surface phenomena to Mediterranean soil conditions, thereby satisfying an important requirement for the successful application of remote sensing techniques (Hill, 1993b; Hill et al., 1995). However, the validity of such concepts has to be carefully analysed in the context of the specific physiographic conditions under which they should be applied. Modifications might be required, for example, in cases of extreme aridity where soil forming processes do not permit the accumulation of noticeable amounts of organic components (e.g., Escadafal, 1994).

2.2.2. *Mapping Vegetation Abundance.*

Vegetation attributes are usually described by structure, dynamics and taxonomic composition, of which taxonomy is the least important of the three. The classification which is most compatible with remote sensing relates to the projected foliage cover (PFC, or cover) and the life form of the tallest vegetation stratum (Graetz, 1990). However, about 75 % of the earth's surface is covered by sparse vegetation which transmits the colour of the soil beneath, i.e. the projected foliage cover (PFC) is below 1. In particular semi-arid ecosystems, such as the Mediterranean, are dominated by such vegetation communities. For these, the soil surface itself should be as much an object of attention as is the vegetation (Graetz, 1990), and the key issue is therefore to provide accurate estimates of green vegetation abundance which are not biased by the spectral contribution of background components (i.e., soils and rock outcrops). Attention should also be given to the spectral characteristics of non-green components of plant canopies and associated plant litters, which largely contribute to the reflectance of terrestrial surfaces in semi-arid ecosystems (Elvidge, 1990). Although we know that the spectral resolution of earth observation satellite systems is not adequate to consistently differentiate between dry plant components and soils, efforts have to be made to resolve ambiguities from the image context (Smith et al., 1990).

3. Advanced Data Interpretation - The Spectral Mixing Paradigm

It results from our conceptual considerations that we require information extraction methods which provide largely unbiased estimates for green vegetation cover, equally permit the identification of soil related spectral information, and allow sufficient standardisation for multi-temporal monitoring. Traditional multi-spectral classification approaches as well as most vegetation indices are not ideally suited to fulfill these requirements (Hill et al., 1995). Since the inversion of physically-based bidirectional reflectance models against satellite data is not easily feasible with currently available data sets, we wish to draw the attention on suitable semi-empirical models.

Uncertainties in measuring vegetation abundance with optical sensors can be minimised by accounting for the reflective properties of sometimes highly variable background materials (e.g., Siegal and Goetz, 1977; Price, 1993; Hill et al., 1995). Conversely, natural vegetation can significantly mask and alter the spectral response of the ground. Therefore, instead of attempting to develop separate soil- or vegetation-related indices which are based on spectral reflectance measurements in single or multiple bands it seems more appropriate to use dedicated spectral decomposition techniques prior to developing thematic indices based on multispectral measurements.

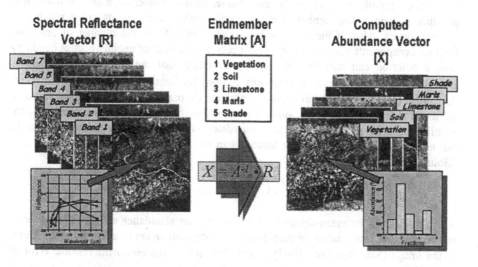

Figure 3. The concept of spectral unmixing, illustrated using an example of a sensor system with six spectral bands and five endmember materials.

One of the most promising approaches for computing the proportional abundance of materials which occur within a specific surface area (i.e., pixel) is based on computationally decomposing multispectral measurements with reference to a finite number of pure spectral components, i.e., endmembers (Figure 3). The method has become known as "Spectral Mixture Analysis" (e.g., Adams et al., 1993; Smith et al., 1990; Craig, 1994), and it assumes that most of the spectral variation in multispectral images is due to mixtures of a limited number of surface materials, and that these mixtures can, in first approximation, be described as a result of additive (linear) spectral mixing (i.e., where each photon contacts only type of surface material). Multiple scattering would need to be either accounted for with non-linear models, or the variables of the linear mixing equation would need to be delinearised by using, for example, single scattering albedo instead of reflectance (Johnson et al., 1983). The computation of proportional abundances can then always be solved with a system of linear equations. In matrix notation one can write

$$A \cdot X = B, \qquad (1)$$

where A is the m · n matrix of spectral endmember signatures (usually derived from so-called spectral libraries, i.e. a suitable collection of laboratory and field measurements), with each column containing one of the endmember spectral vectors. X denotes the n · 1 unknown vector of abundances, and B the m · 1 observed data vector (i.e., measured reflectance of one pixel). The unknown vector of abundances is computed with

$$X = A^{-1} \cdot B \qquad (2)$$

i.e. essentially by inverting the endmember matrix. Evidently, a unique solution is possible as long as the number of spectral endmembers corresponds to the number of spectral bands. Also, if the problem is underdetermined, i.e. the number of unknown fraction components (i.e., abundances) exceeds the number of useful spectral bands by one, a solution can still be obtained by assuming that the set of endmembers is exhaustive[1] (i.e., the sum of the computed endmember fractions is equal to 1). However, the typical case encountered in spectral unmixing (in particular when hyperspectral data are used) is that the number of bands is greater than the number of endmember spectra. The linear mixing model then becomes overdetermined such that the endmember matrix can not be inverted. In this case a solution can be obtained through the pseudo-inverse (Golub and Van Loan, 1983),

$$X = (A^T A)^{-1} A^T \cdot B + \varepsilon \qquad (3)$$

which minimises the mean-squared error in fitting the abundance estimates to the data, and renders the computation of abundance estimates equivalent to a rotational transform of the image (Schowengerdt, 1997). The added term ε represents the residual error of the model.

4. The Assessment of Soil Resources

The mixing paradigm provides a type of image enhancement which is not only intense but also physically meaningful (Craig, 1994). Its objective is of course to isolate the spectral contributions of important surface materials ("endmember abundance") before these are edited and recombined to produce thematic maps (Adams et al., 1989). Hill et al. (1995) have successfully adopted this approach to analyse the spectral information content related to the erosional state of soils (Figure 4), but also to derive precise maps of soil conditions and improved estimates of green vegetation abundance from various types of multispectral images (e.g., Hill, 1993b; Hill et al., 1995).

Spectral mixture analysis allowed them to estimate relative amounts of rock fragments and soil particles on the surface. Since within a specific context, soil erosion leads to an increase in rock cover in source areas and the accumulation of soil material

---

[1] This assumption, however, is not unproblematic since it is difficult to be sure that a sufficient number of spectral endmembers has been defined for a given data set.

as colluvium elsewhere, the content of parent material components could be used as an indicator of degradation. This means that the spectral combination of components is conceptualised in the sense of intimate mixtures (as opposed to spatial-spectral mixing as determined by the sensor spatial response), and corresponds to defining the erosional state of soils as a function of the mixing ratio between developed soil substrates and parent material components which, of course, need to be spectrally distinct from each other. While Fischer (1989) has been adopting these principles for mapping soils of different age in an area of young glacial deposits, Hill et al. (1995) have successfully used the approach to map areas affected from soil degradation and erosion under sub-humid conditions in Mediterranean France (Figure 4).

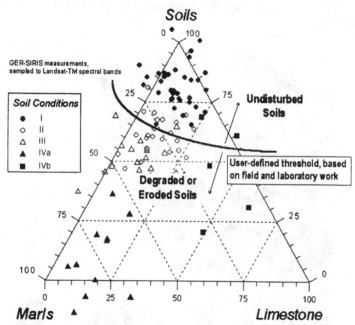

Figure 4. The separation between spectra from undisturbed soils and degraded/eroded substrates, displayed in a ternary diagram of endmember abundances resulting from the unmixing of reflectance spectra with a three endmember model (one soil and two parent material components) (modified after Hill et al., 1995).

TABLE 1. Landsat-based soil condition assignments (combined from 4 individual acquisition dates in 1984) in comparison to 97 field reference locations.

Landsat TM classification	reference locations				number	% correct
	I	II	III	IV		
I	28	9			37	75.7
II	3	13	3		19	68.4
III	1	2	17	6	26	65.4
IV				15	15	100.0
number	32	24	20	21	97	75.3
% correct	87.5	54.2	85.0	71.4	75.3	

142

Thematic maps can then be obtained by applying simple or advanced parametric classifiers to the image of purely substrate-related mixing proportions (i.e., after shade and vegetation contributions were discarded), thereby assigning each pixel to the soil condition class which is closest (Hill et al., 1995). In Figure 5, a simple Euclidian minimum distance classifier was used to produce an Landsat-based soil condition map of the Ardèche Experimental site in S-France. Comparisons to specific ground surveys and available air photos have confirmed that the mapping results are in very good agreement to the spatial patterns of undisturbed and degraded soils in the study site. Additionally, the accuracy of this map was tested with regard to specifically controlled reference locations (Table 1). Taking into account that the geometrical referencing between field locations and their corresponding image position might also affect the accuracy assessment, we consider the precision estimate of 75.3 % an absolutely encouraging result.

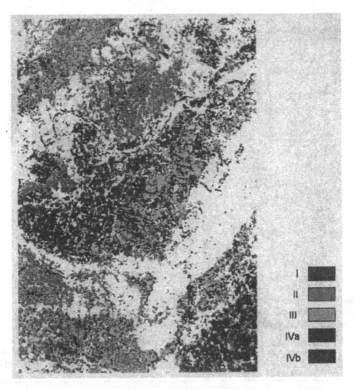

Figure 5. Soil condition map for the Ardèche study site, obtained from the spectral mixture analysis of Landsat-TM data: white areas have an estimated green vegetation cover above 50 % and were not analysed in terms of their soil properties. Please see appendix for image in colour.

Based on hydrological concepts Pickup and Chewings (1988) in Australia developed the erosion cell approach whereby the landscape is divided into the production zone, where there is a net soil loss, a transfer zone with intermittent erosion and deposition and a sink where accumulation occurs. Their approach and the concept of Hill et al. (1995) can also be combined with quantitative estimates of soil properties such as organic carbon, thereby leading to increasingly differentiated possibilities to analyse environmental conditions in dryland ecosystems.

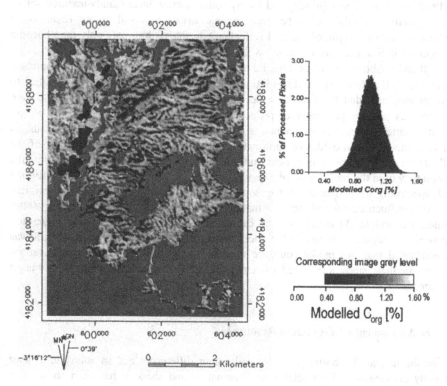

Figure 6. Landsat TM - based map of soil organic carbon in the semi-arid Guadalentin area in SE-Spain. In the image, grey levels correspond to C_{org}-contents between 0.2 and 1.6 %; areas of semi-natural vegetation, irrigated and urbanized land are masked (from Hill and Schütt, in press).

Relating proxy data from a semi-arid ecosystem in SE-Spain (average rainfall 300-350 mm a^{-1}) to the erosion cell approach, Hill and Schütt (in press), for example, suggest that organic carbon in xeromorphic soils can be understood as a tracer substance which highlights areas of accumulation and relative stability (sediment sinks) where soil conditions are favourable because of higher infiltration and water retention capacity, better aggregation, and increased nutrient availability (e.g., Imeson et al., 1996). Based on this premise the objective of differentiating more favourable sink areas from active erosion and transport zones on the basis of optical remote sensing images merely condenses on a detection problem, which is to identify the organic matter content of dryland soils based on spectral indicators. Although the general relationship between organic matter and soil brightness and colour have long been recognised, quantitative relationships are difficult to obtain because albedo is influenced by additional properties such as soil moisture, content of iron, or mafic minerals. However, summarising the results of a review paper by Schulze et al. (1993), one can state that such relationships can be developed within and even among soil landscapes if soil textures are not varying too much. Soil texture also controls whether relationships between soil organic matter and soil colour are linear (in silty and loamy soils) or curvilinear (sandy-textured soils). Consequently, the relationship between organic matter and soil colour (here to be understood as a surrogate of spectral reflectance in the visible) may only then become unpredictable if soil texture varies too widely (Schulze et al., 1993).

Hill and Schütt (in press) proposed a method which can be applied to high resolution spectra as well as to the spectral resolution of Landsat TM, and which can be applied to real images, provided the data have been corrected for atmospheric effects (i.e., are available as primary parameter of reflectance ρ). Also, the specific influence of soil organic matter on spectral reflectance is not expressed in narrow absorption bands, as for example with iron oxides or carbonates, but it determines the overall shape of the reflectance curve in the visible, near and mid-infrared range of the spectrum. For this reason they used the coefficients of a quadratic function (which was fitted to the original soil spectrum) to parameterise the most important characteristics of the original spectrum influenced by soil organic matter, i.e. slope and curvature. Organic matter content (in weight %) could then be estimated through a multiple linear regression between soil organic carbon and the coefficients of the parabolic curve fit; also, it was demonstrated that this model could be used to successfully map the organic carbon content in their study site based on atmospherically corrected Landsat TM images (Figure 6).

5. The Assessment of Vegetation Resources

Given the increased robustness against soil colour differences and ist ability to at least partially compensate effects related to illumination and shade it has been shown that Spectral Mixture Modelling can also provide better estimates of vegetation abundance (i.e., proportional cover) than conventional vegetation indices (Figure 7, see also Lacaze et al., 1995). In particular the incorporation of variable background signals through multiple endmember sets has provided quite promising results (Hill et al., 1995). Whether, and to which extent these methods can be applied to process and interpret long time series of satellite imagery, was meanwhile illustrated in more detail through a case study developed for the island of Crete, Greece. This example is particularly interesting for the context of Albania because most of the countryside is very similar to the mountainous parts of Crete.

Crete is a region which, during the past decades, experienced an increasing pressure on natural resources through excessive grazing activities in mountainous ecosystems, mainly triggered by an increasing demand in animal products. An additional element of the environmental impact related to grazing results from the construction of new access roads which is supported through the Regional Development and Cohesion Fund of the European Union, a funding source which became available after Greece joined the EU in 1981. Areas which traditionally had been too remote, became rapidly accessible by vehicle. Animals, together with additional food supplies, could then be transported into the mountains, with the result that the presence of man and roaming animals became more frequent than ever before. It was therefore believed that a retrospective analysis of vegetation dynamics in the mountainous ecosystems of Crete provides one of the few test cases to assess the evidence of grazing-dependant degradation processes in the Mediterranean (Hill et al., 1998).

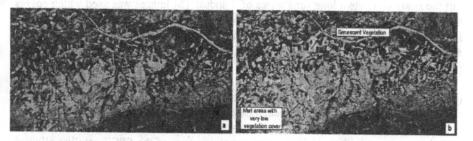

Figure 7. Estimates of green vegetation abundance (i.e., proportional cover), resulting from the use of fixed (a) and spatially adaptive (b) endmember sets in a study site with Mediterranean species, (Ardèche, France) (from Hill et al, 1995). Please see appendix for image in colour.

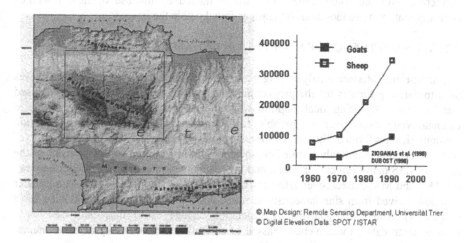

Figure 8. Rangeland monitoring sites in the mountainous ecosystems of Crete, and the increase of ruminants in the Psiloriti region. Please see appendix for image in colour.

The Asteroussia and Psiloriti Mountains of Crete (Figure 8) are hence known for the fact that the grazing pressure has significantly increased during the past decades, in particular since Greece joined the European Communities. Official statistics indicate that the number of sheep and goats in the townships of the Psiloriti region has grown by more than 150 % since 1980 (Zioganas et al., 1998; Dubost, 1998), resulting in "a significant decrease of woody cover and consequently prone to desertification" (Lyrintzis et al., 1998). Notwithstanding the reliability of the statistical figures, it is beyond any doubt that grazing pressure has been continuously high during the past decades. But, has this really triggered a significant loss of biomass, and, if so, does it affect the complete mountain ranges or only specific areas?

5.1. EXTENDED TIME SERIES OF EARTH OBSERVATION IMAGES

For the study on Crete a long time series of Landsat TM images was used which covers the time span between 1984 and 1996. This data set was further extended by adding Landsat-MSS data dating back until 1976. All images were acquired between late May and early June, such that one could assume to deal with information from comparable phenological phases.

In the context of this paper not details are discussed on the pre-processing techniques (i.e. geometric and radiometric corrections) which are required for a precise and standardised assessment of biophysical variables from satellite imagery (see also Lacaze et al., 1996). The Landsat-TM time series was geometrically corrected, where the use of digital elevation data derived from panchromatic images of the SPOT-satellite allowed for the correction of terrain-induced distortions with a high degree of accuracy. A radiative transfer code was employed to atmospherically correct each image, which also included corrections of topography-induced illumination effects. The consistency of the correction over the whole time series was validated by the use of field reference measurements and pseudo-invariant features found within the images .

5.2. MAPPING TRENDS OF VEGETATION DYNAMICS OVER TIME

Linear Spectral Mixture Analysis (Figure 3, see also the discussion in chapter 3) was used to derive estimates for the proportional vegetative cover from each pre-processed Landsat image. By using locally optimised sets of endmember spectra we were able to estimate vegetation cover comparable to the ground-based mapping of vegetation communities performed by the DeMon-II project partners from the University of Iraklion. A correspondingly adapted endmember matrix was used for the analysis of the Landsat-MSS data from earlier years, and the remaining systematic differences between the TM- and MSS-based cover estimates were successfully adjusted based on a transfer function derived from simultaneously acquired TM- and MSS-images from June 1988 (Figure 9). The trend analysis of vegetation change could thus be based on a homogeneous data set which encompassed 13 carefully calibrated and processed images, covering a time span from 1976 to 1996.

One of the difficult problems encountered in analysing time series of vegetation abundance estimates is to differentiate between single events (resulting either from phenological pulses, or from abrupt land use changes) and true long-term trends. Similar to climate change studies, we have tried to use a regression approach for characterising the long-term trends resulting from the satellite-derived time series of proportional vegetative cover. Although the coefficients of determination for the regression at most pixel positions are low, the approach provides significant and useful information. Some selected examples may illustrate how long-term changes are depicted (Figure 10). The results obtained for one of the last contiguous mature Kermes oak forests in the Psiloriti mountains (Rouvas Forest), for example, demonstrate that, apart from some minor variations, the vegetation cover/abundance during the last 20 years has been quite stable, since moderately negative or positive regression slope are not considered to represent a significant trend. The extended grasslands of the Nida plain exhibit a similar degree of stability (which perfectly conforms to field observations), but have much stronger inter-annual variations to be straightforwardly explained by phenological shifts. In comparison, areas truly affected by long-term vegetation losses are characterised by noticeably steeper trend lines (as they appear, for example, in some areas of the Kouloukonas and Psiloriti Mountains).

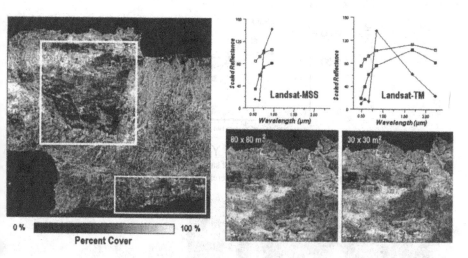

Figure 9. Proportional Vegetative Cover (in %) in the Crete study sites, derived through Spectral Mixture Analysis from the 1988 Landsat-TM scene (left). Landsat-MSS and Landsat-TM data differ in spatial and spectral resolution, and require correspondingly adjusted endmember spectra (top right). Note that, due to the applied atmospheric and topographic corrections, no relief-induced shading is perceived in the images.

The analysis of temporal trajectories from such reference sites is an important prerequisite for defining a set of suitable threshold values which can be applied to interpret the regression-based trend analysis of complete images. In the Psiloriti highlands, for example, negative trends are only identified in specific areas, while large parts of the region must be considered stable, or even exhibited increasing values of vegetation cover (Figure 11). A more detailed analysis of the results reveals that the summit regions, but also most of the severely rugged (and thus not easily accessible) highland sites, are found within the stable areas.

148

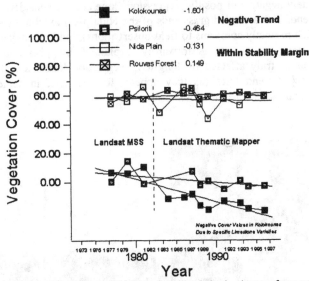

Figure 10. Regression-based trend analysis of satellite-derived estimates of proportional vegetative cover for selected reference areas in central Crete.

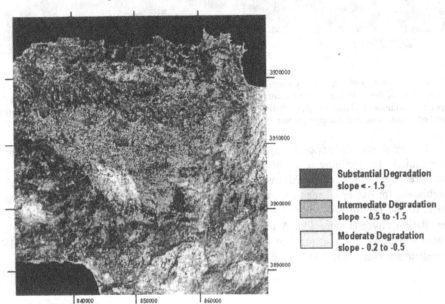

Figure 11. Trend analysis of vegetation cover for the period 1977-1997, based on Landsat-MSS and -TM time series covering the Psiloriti highlands above 1000 m. The colours indicate areas where degradation processes led to a significant loss of perennial plant cover; areas with stable (or partially increasing) vegetation cover are not colour-coded. Please see appendix for image in colour.

Summarising the preliminary results referred to in the previous section, it is concluded that the retrospective assessment of vegetation dynamics has been able to cast some doubt on frequently cited statements that, starting with the EU membership of Greece, most of the mountainous ecosystems in Crete have been entering a phase of accelerated degradation triggered by intensified grazing. Of course, herding has been encouraged by European Community subsidies and, although the official figures seem somewhat inflated, the number of browsing animals has certainly increased (e.g. Rackham and Moody, 1996). However, based on sufficiently long time series of Earth Observation Satellite imagery, it could here be demonstrated that degradation trends are in fact restricted to specific areas, while extended parts of the central Crete highlands remained largely unaffected. This appears to be in good agreement with the fact that most of the increasing grazing activities are, and have been, relying on the use of access roads for transporting animals and additional food supplies. It is thus not surprising that we find the most extended areas with long-term depletion of vegetation in the accessible highlands which extend north and north-east of the Psiloriti summit, and within the Kolokounas range (Figure 11). Vegetation dynamics for most of the remaining areas in Asteroussia and Psiloriti exhibit patterns typical for a steady state equilibrium, as they are known from large areas in the Mediterranean which, for several decades, have been excluded from land use pressure.

6. Conclusions

The advantages of satellite remote sensing result from its synoptic nature, comprehensive spatial information and objective, repetitive coverage. It has been illustrated how remotely sensed primary parameters, such as the spectral surface reflectance ρ, can be converted into a standardised characterisation of soil conditions and vegetation abundance. In this context, the term of thematic concepts has been introduced, by which we understand the conceptual background for identifying functional links between surface reflectance and vegetation and soil characteristics. Such concepts are primarily based on research in geosciences and ecology, and it is important to keep strong links between these disciplines and remote sensing specialists. While remote sensing has initially been used primarily for resource mapping and inventory it turns out that monitoring and predictive modelling is becoming more important and successful. However, no matter whether one intends to map surface properties or the objective is to estimate fluxes based on remote sensing data requires that the primary parameters, such as the spectral reflectance or surface temperature, are first retrieved with adequate precision. This raises immediately the issue of adequate radiometric corrections. While atmospheric effects have been a significant factor in the failure of scene models (in particular for monitoring concepts) much progress has been achieved in the past years, and we are now in the position that large parts of the radiometric pre-processing are considered a routine operation, similar to the geometric rectification. Presently, the remaining problems in retrieving surface reflectance from satellite data appear more related to the absolute radiance calibration of the sensor systems than to methodological drawbacks. This ensures that more advanced scene models, such as the spectral mixing paradigm or invertible physically-based analytical models, can be used to derive quantitative estimates and improved indicators for land resources and degradation processes.

Concerning the case studies presented in this paper, a synoptic interpretation of all processed data has allowed to draw first conclusions concerning the dynamic development of selected Mediterranean vegetation communities. The fact that it has been possible to identify a degradational reduction of vegetative cover in specific locations with continued grazing pressure, implies that remote sensing can provide substantial contributions to a more conscious management of precious lands: available resources are limited, and man has already frequently crossed sensible thresholds without taking note in time. It is believed that thorough assessments of available resources, the implementation of adequate management strategies and efficient approaches to monitor and control the state of the environment are core elements on which to build efficient strategies to mitigate land degradation and desertification risks.

We have shown that both inventory and mapping is required to define the current status of soil and vegetation resources and provide a baseline for monitoring, and that surveys must be repeatable and comparable, and therefore demand a standardised methodological framework. Although it is unrealistic that remote sensing will replace traditional sources of data for inventory and monitoring there is, without any doubt, an obvious role it has to play in assessing and monitoring the state of the environment. It thus forms the basis for drafting and implementing efficient land management plans which are needed to avoid land degradation under inadequate management practises.

7. References

1. Adams, J.B., M.O. Smith and A.R Gillespie (1989) Simple models for complex natural surfaces: a strategy for the hyperspectral era of remote sensing, *Proc. of the IGARSS '89 Symposium, July 10-14, Vancouver*, vol. 1, 16-21.

2. Adams, J.B., M.O. Smith and A.R Gillespie (1993) Imaging spectroscopy: interpretation based on spectral mixture analysis, in C.M. Pieters and P.A.J. Englert, (eds.), *Remote geochemical analysis : elemental and mineralogical composition*, Cambridge University Press: Cambridge, pp. 145-166.

3. Baumgardner, M.F., E.R. Stoner, L.F. Silva, and L.L. Biehl (1985), Reflective properties of soils, in: N. Brady, (ed.), *Advances in Agronomy*, 38, Academic Press, New York, pp. 1-44.

4. Birkeland, P.W. (1990), Soil-geomorphic research - a selective overview, *Geomorphology*, 3, 207-224.

5. Craig, M.D. (1994) Minimum-volume transforms for remotely sensed data, *IEEE Transactions on Geoscience and Remote Sensing*, 32, 542-552.

6. Dubost, M. (1998) European politics and livestock grazing in Mediterranean ecosystems, in V.P. and D. Peter (eds.), *Ecological basis of livestock grazing in Mediterranean ecosystems*, EUR 18308 EN, Office for Official Publications of the European Communities: Luxembourg, pp. 298-311.

7. Elvidge, C.D. (1990) Visible and near infrared reflectance characteristics of dry plant materials, *Int. J. Remote Sensing*, 11, 1775-1795.

8. Escadafal, R. (1993), Remote sensing of soil color: principles and applications, *Remote Sensing Reviews*, 7, 261-279.

9. Escadafal, R., A. Belgith, and A. Ben Moussa (1994), Indices spectraux pour la télédétéction de la dégradation des milieux naturels en Tunisie aride, Proc. 6th Int. Symp. on „Physical Measurements and Signatures in Remote Sensing", 17-24 January 1994, Val d'Isere, France, CNES, Paris, pp. 253-259.

10. Fischer, A.W. (1991) Mapping and correlating desert soils and surfaces with imaging spectroscopy, *Proc. of the 3rd Airborne Visible/Infrared Imaging Spectrometer (AVIRIS) Workshop*, JPL Publication 91-28, Pasadena, 23-32.

11. Graetz, R.D. (1990) Remote sensing of terrestrial ecosystem structure: an ecologist's pragmatic view, in R.J. Hobbs and H.A. Mooney, (eds.), *Remote sensing of biosphere functioning*, Springer-Verlag: New York, pp. 5-30.

12. Graetz, R.D. (1996), Empirical and practical approaches to land surface characterisation and change detection, in J. Hill and D. Peter (eds.), *Remote sensing for land degradation and desertification monitoring in the Mediterranean basin, State of the art and future research*, EUR 16732 EN, Office for Official Publications of the European Communities: Luxembourg, pp. 9-21.

13. Grenon, M., and M. Batisse, eds. (1989) *Futures for the Mediterranean basin, the Blue Plan*, Oxford University Press: Oxford.

14. Hill, J. (1993a) High precision land cover mapping and inventory with multi-temporal earth observation data. The Ardeche experiment, EUR 15271, Office for Official Publications of the European Communities: Luxembourg.

15. Hill, J. (1993b) Monitoring land degradation and soil erosion in Mediterranean environments, *ITC Journal*, 4, 323-331.

16. Hill, J., J. Mégier, and W. Mehl (1995), Land degradation, soil erosion and desertification monitoring in Mediterranean ecosystems, *Remote Sensing Reviews*, 12, 107-130.

17. Hill, J., and D. Peter, eds. (1996) The use of remote sensing for land degradation and desertification monitoring in the Mediterranean basin. State of the art and future research, EUR 16732 EN, Office for Official Publications of the European Communities: Luxembourg.

18. Hill, J., P. Hostert, G. Tsiourlis, P. Kasapidis, and Th. Udelhoven (1998) Monitoring 20 years of intense grazing impact on the Greek island of Crete with earth observation satellites, *J. Arid Environments*, 39, 165-178.

19. Hill, J., and B. Schütt, in press, Mapping complex patterns of erosion and stability in dry Mediterranean ecosystems, *Remote Sensing of Environment*.

20. Imeson, A.C., Pérez-Trejo, F. and Cammeraat, L.H. (1996) The response of landscape units to desertification. in C.J. Brandt and J.B. Thornes (eds.), *Mediterranean desertification and land use*, John Wiley and Sons: Chichester, pp. 447-469.

21. Johnson, P.E., M.O. Smith, S. Taylor-George, and J.B. Adams (1983) A semi-empirical method for analysis of the reflectance spectra for binary mineral mixtures, *J. Geophys. Res.*, 88, 3557-3561.

22. Johnston, R.M., and M.M. Barson (1990), *An assessment of the use of remote sensing techniques in land degradation studies*, Bureau of Rural Resources, Bulletin No. 5, Australian Government Printing Services; Canberra.

23. Kramer, H.J. (1996) Observation of the earth and its environment: survey of missions and sensors, Springer-Verlag: Berlin.

24. Lacaze, B., J. Hill, and W. Mehl (1995), Evaluation of green vegetation fractional cover in Mediterranean ecosystems from spectral unmixing of Landsat-TM and AVIRIS data, in E. Mougin, K.J. Ranson, and J.A. Smith (eds.), *Multispectral and Microwave Sensing of Forestry, Hydrology and Natural Rresources*, Proc. SPIE 2314, 339-346.

25. Lacaze, B., V. Caselles, C. Coll, J. Hill, C. Hoff, S. de Jong, W. Mehl, J.F.W. Negendank, H. Riezebos, E. Rubio, S. Sommer, J. Teixeira Filho, and E. Valor (1996), *Integrated Approaches to Desertification Mapping and Monitoring in the Mediterranean Basin. Final Report of the DeMon-1 Project*, edited by J. Hill, EUR 16448 EN, Office for Official Publications of the European Communities: Luxembourg.

26. Lyrintzis, G., V.P. Papanastasis, and I. Ispikoudis (1998) Role of livestock husbandry in social and landscape changes in White Mountains and Psilorites of Crete, in V.P. Papanastasis and D. Peter (eds.), *Ecological basis of livestock grazing in Mediterranean ecosystems*, EUR 18308 EN, Office for Official Publications of the European Communities: Luxembourg, pp. 322-327.

27. Perez-Trejo, F. (1994) *Desertification and land degradation in the European Mediterranean*, EUR 14850 EN, Office for Official Publications of the European Communities: Luxembourg.

28. Pickup, G., and V.H. Chewings (1988) Forecasting patterns of soil erosion in arid lands from Landsat MSS data, *Int. J. Remote Sensing*, 9, 69-84.

29. Price, J.C. (1993) Estimating leaf area index from satellite data, *IEEE Transactions on Geoscience and Remote Sensing*, 31, 3, 727-734.

30. Rackham, O., and J. Moody (1996) *The making of the Cretan landscape*, Manchester University Press: Manchester.

31. Schulze, D.G., J.L. Nagel, G.E. Van Scoyoc, T.L. Henderson, M.F. Baumgardner, and D.E. Stott (1993) Significance of organic matter in determining soil colours, in J.M. Bigham and E.J. Ciolkosz (eds.), *Soil Color*, Soil Science Soc. of America: Madison, Wisconsin, pp. 71-90.

32. Schowengerdt, R.A. (1997) *Remote sensing. Models and methods for image processing*, Academic Press: San Diego.

33. Siegal, B.S., and A.F.H. Goetz (1977) Effect of vegetation on rock and soil type discrimination, *Photogrammetric Engineering and Remote Sensing*, 43, 191-196.

34. Smith, M.O., S.L. Ustin, J.B. Adams, and A.R. Gillespie (1990) Vegetation in deserts: I. A regional measure of abundance from multispectral images, *Remote Sensing of Environment*, 31, 1-26.

35. Thomas, D.S.G., and N.J. Middleton (1994) *Desertification. Exploding the myth*, John Wiley & Sons: Chichester .

36. UNEP (1991) Status of desertification and implementation of the United Nations plan of action to combat desertification, UNEP: Nairobi.

37. Verstraete, M.M. (1994), Scientific issues and instrumental opportunities in remote sensing and high resolution spectrometry, in J. Hill and J. Mégier (eds.), *Imaging spectrometry - a tool for environmental observations*, Kluwer Academic Publishers: Dordrecht, pp. 25-38.

38. Zioganas, C., E, Anephalos, and V.P. Papanastasis (1998) Livestock farming systems and economics in the Psilorites mountain of Crete, Greece, in V.P. Papanastasis and D. Peter (eds.), *Ecological basis of livestock grazing in Mediterranean ecosystems*, EUR 18308 EN, Office for Official Publications of the European Communities: Luxembourg, pp. 328-331.

COASTAL ZONE GEOMORPHOLOGICAL MAPPING USING LANDSAT TM IMAGERY: AN APPLICATION IN CENTRAL ALBANIA

P. CIAVOLA, U. TESSARI, F. MANTOVANI, M. MARZOTTO, U. SIMEONI.
Dipartimento di Scienze Geologiche e Paleontologiche
Università degli Studi di Ferrara,
Corso Ercole I d'Este 32
44100 – Ferrara
Italy

Introduction

Albania is country with a rugged landscape (Shqiptar, country of eagles), with plains corresponding to only 1/12 of the whole national surface, mostly located along the coastline. The coastline has a total length of about 380 km, of which 284 km are along the Adriatic Sea, while the rest is facing the Ionian Sea. The last population census (1990) indicates that the workforce is 44% of the population (3 256 000 inhabitants), mainly employed by the state (63%) with the rest of the workers organised in co-operatives or working in agriculture. Agriculture was traditionally one of the main economical activities, able to satisfy 93% of the national demand; it has been declining from the late 1980s onwards, due to migration of farmers towards the main towns, abandoning traditional agricultural practices.

At the end of the Second World War the arrival of the communist regime was followed by large-scale reclamation projects to convert wetlands into arable land. Land reclamation was undertaken in parallel with river engineering of the Drini, Mati, Semani and Vjosë, together with the construction of large reservoirs. This paper will present an evaluation of the impacts that these large engineering projects caused on the landscape, putting changes occurred in the last 20 years into the context of high variability that is typical of the Albanian coastline [13].

Although such an evaluation could have been carried out using traditional methods such as ground mapping or cartographic analysis, the recent technological advances in satellite remote sensing, digital mapping and geographic information systems offer an opportunity to cover large areas with a good level of detail (e.g. 30 m pixel resolution for the Landsat TM, bands 1-5), providing that some ground truthing is carried out. The interpretation of satellite images was integrated with field visits to the area in 1994 [3]. The method of using remotely sensed data for geomorphological mapping of deltaic areas is well established, previous application were those of [5, 9, 11, 12].

M.F. Buchroithner (ed.),
Remote Sensing for Environmental Data in Albania: A Stragegy for Integrated Management, 153–163.
© 2000 *Kluwer Academic Publishers. Printed in the Netherlands.*

154

1. The Coastal Plain of Myzeq: A Case Study

The case study area (Figure 1) presented here is the largest coastal plain of Albania (about 1900 km^2) that includes the floodplains of three of the largest rivers of this territory, the Shkumbini, the Semani and the Vjosë. These three rivers account for 49 % of the water discharge from the Albanian mainland into the Mediterranean Sea [10]. Two of the largest lagoons of the country are present in this area, Karavasta (42 km^2), between the deltas of the Shkumbini and Semani, Nartes (41 km^2) between the Semani and Vjosë. This stretch of coastline, with a microtidal range, includes a variety of geomorphological environments such as deltas, coastal wetlands, dunes, beaches and beach ridges. The buiding up of the wide coastal plain during the Holocene was controlled by rapid accretionary events in delta growth [9] and ancient coastal towns such Apollonia, born 2500 years ago 5 km away from the coast, are now located 8 km inland.

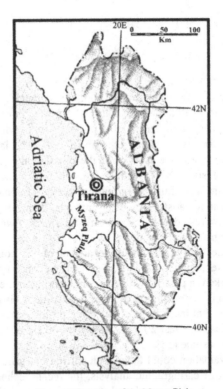

Figure 1. Location of the Myzeq Plain.

The evolution of the Myzeq plain has been characterised by rapid coastal change: river courses became abandoned on a time scale of decades, while new ones were created, leaving behind a strandplain with marshes and wetlands. While in the present century man has played an important role diverting rivers and building dams [4, 13] the early

Holocene evolution of the coastal zone was probably affected by large subsidence generated by neotectonic faulting [9].

2. Data Sources

This study used images captured by the Landsat 5 Thematic Mapper on 18 October 1986 (Figure 2) and on 29 October 1996 (Figure 3). Each scene covers an area of 185×185 km^2 and the bands of the sensor used for geomorphological mapping were Band 1 (0.45-0.52 µm) and Band 5 (1.55-1.75 µm); the first one (blue) provided information on movement of water masses in the nearshore area and soil/vegetation differentiation, while the second one (mid IR) supplied information on vegetation moisture, expressed in the imagery as changes in tones, from light grey towards black, due to a variation in reflectance values.

The visual examination of the images in Band 5 outlines a variation in tones and patterns: in the 1986 scene the reflectance values are higher, tones are lighter and the field patterns are well defined; in the 1996 image there is a predominance of dark tones, a sign of high humidity in soils and the patterns of the fields are not seen so clearly. This is probably related to a lack in maintenance of drainage works in the area due to the political change that occurred in the country in the early 1990s.

The images were imported into a Geographical Information System (IDRISI for Windows) and georeferencing was carried out using ground control points identified on the 1979 topographic base of the Geological Map of Albania at a 1:200 000 scale [7]. To aid in the photointerpretation, the bands of the two scenes were combined to obtain a false colour composite, assigning a red colour to the 1986-Band 5, a blue colour to the 1996-Band 1 and a green colour to the 1996-Band 5. The obtained colour composite is presented in Appendix at the end of this volume. In order to identify areas where drainage variations had occurred, Band 5 from the 1996 scene was subtracted from the equivalent in the older image. The subtraction between equivalent pixels produced a third image, where in white are indicated areas of land gain, e.g. areas of accretion along the coastline, and in black areas of land loss, e.g. saturated soils due to a shallow water table and/or flooding, using the fuzzy number method within the IDRISI Software [8]. The insets in the image in Appendix are an example of the application of this technique to the main deltaic areas.

TABLE 1. Cartography and orthogonal aerial photography used in the study.

Data Source	Author	Scale
Mittel Albanien map 29	Austrian Navy (1870)	1:100 000
Aerial Photographs 3/NA/121	UK Royal Air Force (1943)	1:10 000
Valona-Semani nautical chart	USSR Navy (1955)	1:100 000
Geological map of Albania	Albanian Geological Institute (1982)	1:200 000

156

The two coastlines and the different geomorphological units presented below were digitised on screen within the GIS. In order to identify the age of land reclamation or river diversion events, a series of cartographic sources and aerial photographs were used (Table 1).

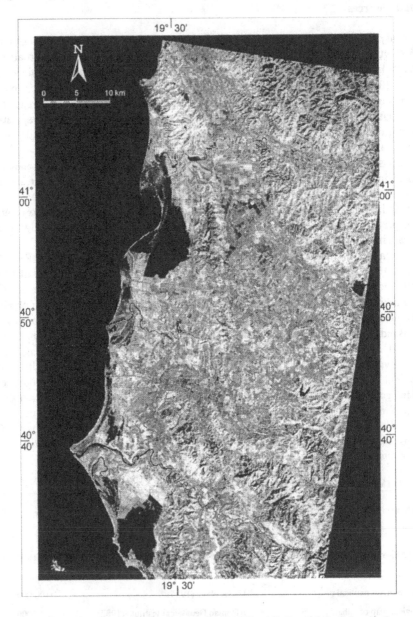

Figure 2. Landsat TM image captured on 18 October 1986 (Band 5)- Copyright by Telespazio S.p.a.

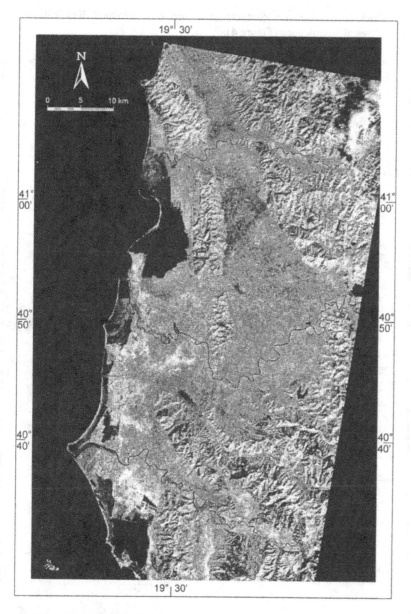

Figure 3. Landsat TM image captured on 29 October 1996 (Band 5)- Copyright by Telespazio S.p.a.

3. Geomorphological Mapping

The mapping criterion that was used is based on identifying on screen the photographic and morphological characteristics of each geomorphological element.

Objects which have common characteristics are then grouped into categories of geomorphological significance (Figure 4).

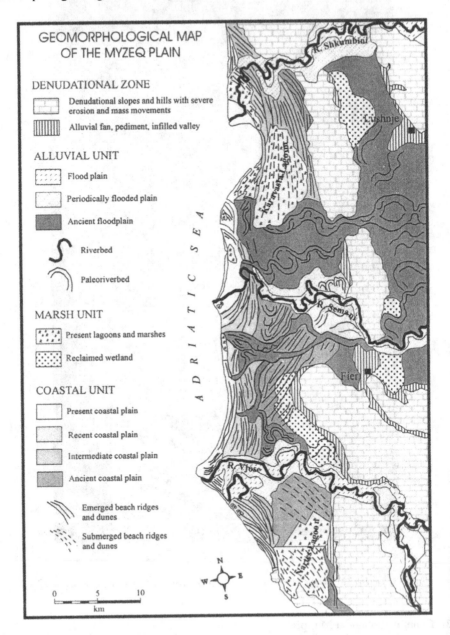

Figure 4. Geomorphological map of the area.

3.1. COASTAL UNITS

The Coastal Units cover an area of 232 km^2 where natural textures are predominant, mainly related to coastline migration. The main landforms are sandy beaches, systems of coastal dunes (with or without vegetation) and small lagoons. Lineations have a concave shape seawards; their orientation, together with beach surveys undertaken during the field work period, allowed a sub-classification into four different units, corresponding to different phases of coastal accretion. In proximity of the deltas the lineations related to beach ridges are interdigitated with those related to delta growth. In the area of Karavasta, the lineations corresponding to beach ridges of the same age can be followed at each side of the lagoon, indicating that the lagoon formed after the period of coastal accretion that created the ridges. Mathers *et al.* [9] reached the conclusion that the lagoon formed due to neotectonic processes in the last 1 000 years; however, they do not present in their paper any dating evidence. The previous authors draw a tectonic alignment, e.g. a hinge line or a fault along the present barrier system separating Karavasta from the open sea. According to them, the tectonic alignment can be followed as far south as the western edge of Nartes Lagoon. The patterns of land use follow the lineations described above and are generally parallel to the coastline.

3.2. COASTAL WETLANDS

These units cover a total surface of 423 km^2 and have as distinctive characteristics dark tones in the images, indicating high levels of soil humidity. It is worth to notice that in the deltaic areas there is a general increase in the dark tones in the more recent imagery, indicating either flooding or a shallow water table. To notice that in most of the Shkumbini and Semani fieldsites that were visited in 1994, many abandoned pumping stations were observed. The reason for such hydrological characteristics has to be seen in the present agricultural usage of many former wetlands, that were reclaimed during the communist regime. The origin of a reclaimed wetland can be identified using the land use pattern. Saltmarshes and back-barrier lagoons are generally reclaimed and reworked following the coastline orientation or orthogonal to it. On the other hand, field patterns along the rivers follow changes in meander shape, therefore are less aligned that those in the coastal areas. Another element of distinction between reclaimed salt and freshwater wetlands is the field size. In coastal areas the fields have regular patterns and constant sizes: infrastructures like roads and ditches follow the orientation of the unit patterns, indicating that both reclamation and development occurred at the same time. Reclaimed freshwater marshes have generally a land use with small fields and a change in pattern orientation can be seen across the images, since the farmers are forced to shift their crops according to flood events. Besides, field inspection identify a not very productive sandy soil in the coastal area (e.g. the southern edge of Karavasta), due to lack of water and presence of salts, while at the rivers mouths the crops are healthier due to the fertility of the soil and water availability. The integration of Landsat images with historical sources identified two major reclamation phases: up to the early 1950s a total of 154 km^2 of wetlands were reclaimed (42 km^2 of saltmarshes and 112 km^2 of freshwater marshes); the second reclamation phase (1950s-onwards) created 269 km² of land to the expense of 186 km² of saltmarshes and 83 km² of freshwater wetlands.

3.3. RIVERINE UNITS

These cover most of the coastal plain (452 km²) and consist of corridors that follow the three rivers. Using as diagnostic element the presence of man-made patters (fields), two sub-units were distinguished: the 'active' floodplain and the palaeo-floodplain, separated by an intermediate area which is periodically flooded only during exceptional events. The active fluvial units correspond to the lowest order of river terraces, covering a total of 7.12 km² along the Skumbini, 34.72 km² along the Semani and 29.68 km² along the Vjosë. They are often delimited by canals and roads and farming is not too successful, probably because of the frequent floods in the winter period, that are typical of the area [4, 10]. The degree of exploitation of these units by farmers increases as we move away from the river. Abandoned meanders can still be seen within this unit and small oxbow lakes act as reservoirs for irrigation.

3.4. DENUDATION UNITS

The geomorphological characteristics of these units were not examined in a great level of detail because they are beyond the scope of this paper which essential deals with alluvial sedimentation in the coastal plain. They correspond to rock outcrops along the hills that border the plain, which consist of Pliocene sandstones and pebbly sandstones [7]. Slope erosion and landslides are present and pediment deposits are seen at the heads of fluvial valleys.

4. Delta Evolution

Delta shape results from the interaction between sediment discharge by the river, wave energy and tidal effects [15]. Taking these factors into consideration, it is possible to locate the three deltas of the study region within the traditional tripartite diagram proposed by Galloway [6] as in Figure 5.

Detailed cartographic studies presented by previous authors [2, 4, 10] discovered that the morphology of these deltas changes several times in the last 120 years. The Vjosë delta is the only one that maintained a constant morphology throughout this period, typical of a wave-dominated environment, while the Shkumbini has a morphology that indicates a more consistent sediment discharge. In the first 50 years of the 20th century the Semani shifted its mouth northwards from the position it occupied in the late 19th century, at a speed of 200 m yr⁻¹, changing from a wave dominated morphology to a digitated one (Figure 5), river-dominated. Later the Semani migrated back towards its previous position and form, following floods that occurred in the mid 1960s. All the previous studies agree that there was a decrease in sediment discharge in the period 1960s-1970s, related to river diversion, reclamation works in the floodplain, construction of dams and dredging of river beds.

The landscape classification of the coastal plain that was presented in the previous paragraphs obviously reflects these river dynamics. Ciavola *et al.* [4] reviewed the hydrological characteristics of these three rivers, comparing their catchment surface and sediment yield with worldwide examples. The rivers of Albania have large sediment loads and their high sediment yields, which can be considered a proxy of sediment transport effectiveness, imply large sediment inputs into the coastal zone, despite the relatively small water discharges. Most of the sediment is coarse, since Pano [10] estimates bed-load percentages of 16-23%, but is at present trapped (up to 70%) within dams built in the upper river course.

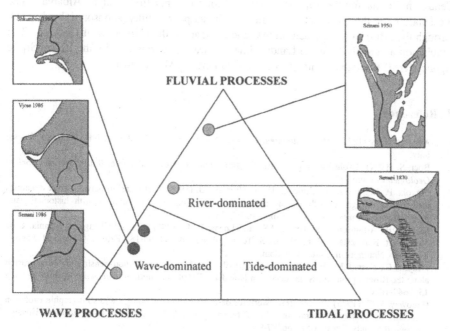

Figure 5. Classification of the deltas of Semani, Skumbini and Vjosë using the tripartite diagram of Galloway (1973)

5. Conclusions

The study presented in this paper has proved that Remote Sensing can be used as a simple tool for geomorphological mapping of the coastal zone of Albania. Clearly it cannot replace traditional topographic or geological surveying but has the main advantage of a possible use of imagery repeated through time. The studied coastal plain resulted controlled by river processes and through a large extend by river engineering works. The impact of a decrease in artificial drainage of reclaimed wetlands was also evaluated together with land-use changes. The studied flood-plain testifies that Albanian rivers are extremely dynamic and future development plans should take this into account. Remotely sensed images introduced into a GIS, together with detailed topographic maps of the floodplains could be used to produce Digital Elevation Models for planning purposes. The integration of DEM with statistical data (maximum level and return period) of significant flood events could then be used to produce risk assessment maps, to aid in the development of land corridors along the rivers.

6. Acknowledgements

The first author acknowledge Prof. E. Samimi and the staff at the Geographical Study Centre in Tirana for the support given to him since his first visit to Albania. The presentation of the paper at the Tirana Workshop was fully sponsored by NATO, through the effort of Prof. Buchroithner and his team at the University of Dresden. The acquisition and analysis of the Landsat imagery was made possible by the Italian Space Agency (ASI) through a contract to one the authors (F. Mantovani).

7. References

1. Austrian Navy (1872) *Mittel Albanien*, Coastal Map n. 29, scale 1:100,000, F.H. Schimpff, Trieste Italy.
2. Boçi, S. (1994) Evoluzione e problematiche ambientali del litorale albanese, *Bollettino della Società Geologica Italiana* 113, 7-14.
3. Ciavola, P., Mantovani, F., Simeoni, U., and Tessari, U. (1999) Relationship between river dynamics and coastal changes in Albania: an assessment integrating satellite imagery with historical data, *International Journal of Remote Sensing* 20, 561-584.
4. Ciavola, P., Arthurton, R.S., Brew, D.S., and Lewis, P.M. (1995) Coastal Change in Albania: Case Studies at Karavasta and Patok, BGS Technical Report WC/95/18, British Geological Survey, Keyworth, Nottingham, United Kingdom.
5. Frihy, O.E., Nasr, S.M., El Hattab, M.M., and El Raey, M. (1994) Remote sensing of beach erosion along the Rosetta promontory, northwestern Nile delta, Egypt, *International Journal of Remote Sensing* 15, 1649-1660.
6. Galloway, W.E. (1975) Process framework for describing the morphologic and stratigraphic evolution of deltaic depositional systems, in: M.L. Broussard (ed.), *Deltas, models for exploration*, Houston Geological Society, Texas, USA, pp. 87-98.
7. Instituti I Studimeve dhe I Projektimeve tè Gjelogjise (1983) *Geological map of Albania*, scale 1: 200,000, Department of Energy and Mines, Tirana, Albania.
8. Marzotto, M. (1998) Analisi geomorfologica ed ambientale della fascia costiera albanese tra Durazzo e Valona, Unpublished master thesis in environmental sciences, University Cà Foscari of Venice, 157 pp.
9. Mathers, S., Brew, D.S., and Arthurton, R.S. (1999) Rapid Holocene evolution and neotectonics of the Albanian Adriatic coastline, *Journal of Coastal Research* 15, 345-354.
10. Pano, N. (1992) Dinamica del litorale Albanese (sintesi delle conoscenze), *Proceedings of 19th A.I.G.I. Meeting*, G. Lang Publishing, Genova, pp. 3-18.
11. Pramanik, M.A.H., Ali, A., Quadir, D.A., Rahman, A., Shadid, M.A., and Hossain, M.D. (1987) Conference report on the regional seminar on the application of remote sensing techniques to coastal zone management and environmental monitoring, Dhaka, Bangladesh, 18-26 November 1986, *International Journal of Remote Sensing* 8, 659-673.
12. Sgavetti, M. and Ferrari C. (1988) The use of TM data for the study of a modern deltaic depositional system, *International Journal of Remote Sensing* 9, 1613-1627.
13. Simeoni, U., Pano, N., and Ciavola, P. (1997) The coastline of Albania: morphology, evolution and coastal management issues, in: F. Briand and A. Maldonado (eds.), *Evolution des côtes méditerrannéennes*, Volume 3 Science Series, Commission Internationale pour l'Exploration Scientifique de la mer Méditerrannées, Monaco, pp. 151-168.

14. USSR Navy (1955) *From the bay of Valona to the River Semani*, nautical chart scale 1:100,000, VMS Maritime Editions, Moscow, CIS.
15. Wright, L.D., and Coleman, J.M. (1973) Variations in morphology of major river deltas as functions of ocean wave and river discharge regimes, *American Association of Petroleum Geologists Bulletin* **57**, 370-398.

SNOW RUNOFF MODELS USING REMOTELY SENSED DATA

EBERHARD PARLOW
Institute of Meteorology, Climatology and Remote Sensing
University Basel
Spalenring 145
CH-4055 Basel

Abstract

The paper shows how satellite data can be used to analyse snow coverage in mountainous areas and to use satellite data to drive models of snow melt processes. Snow melt models are mostly based on a degree-day-factor approach which is taken to parameterise energy uptake of the snow cover for melting. Another approach are physically based models with a complete description of the radiation and heat exchange between the snow cover and the boundary layer of the atmosphere including the effects which are related to topography (altitude, slope and aspect) and land cover. An examples of snow coverage classification using a degree-days-approach is given from the Swiss Alps. The analysis of the spatially distributed net radiation as the driving energetic factor of snow melt processes by using satellite data of various sensors (NOAA-AVHRR and LandSat-TM) is documented from a catchment area of the arctic archipelago of Spitsbergen.

1. Introduction

In many regions of the world water supply is an important aspect for environmental management strategies. Reasons for a better knowledge about water resources are:

- Water management
- Flood control and hazard prevention
- Agricultural management
- Hazard prevention
- Hydro-power supply
- Regional planning and sustainable development of regions

Especially in mountain areas snow accumulation and snow melting are highly variable in space and time due to topographic conditions. This problem increases in regions with a distinct dry season like the Mediterranean winter rain subtropics, where the winter accumulation of snow is an important ecological factor for the regional water supply during spring and early summer. Normally the network of measurements is not

165

M.F. Buchroithner (ed.),
Remote Sensing for Environmental Data in Albania: A Stragegy for Integrated Management, 165–178.
© 2000 *Kluwer Academic Publishers. Printed in the Netherlands.*

sufficient to collect enough data as input for numerical models. Satellite remote sensing with data from NOAA-AVHRR, SPOT, IRS-LISS or LANDSAT-TM platforms can help to bridge the gap between widely distributed measurements of snow depth, snow density or precipitation and the spatially distributed information needed to run numerical models properly.

2. Data

There is a variety of different satellite systems available which can be used to monitor snow coverage in mountainous regions. From the technical point of view they can be subdivided into a) sensors with medium spatial and high temporal resolution, b) sensors with low temporal but high spatial resolution and c) microwave systems. The choice of satellite data is depending on a number of variables: in polar/sub-polar regions the limitation of sunlight during winter implies nearly automatically the choice of radar sensors out group 1 in Table 1. If there are frequent changes in snow coverage or high cloud coverage which reduces visibility of the land surface from satellite, a high temporal resolution is prerequisite and satellite data from group 2 (Table 1) should be used. For local analysis or when topographic features are very complex like in high alpine terrain the integration of high resolution satellite data could be appropriate, even if the repetition cycle is in a range of two to three weeks. Very good results were achieved when data fusion of various sensors was carried out.

TABLE 1. Selection of satellite sensors which can be used for snow cover detection.

Satellite	Sensor	Spatial resolution (m)	Swath width (km)	Temporal resolution (days)	Bands	Spectral range (μm)
Group 1						
ERS-2	ATSR2	1000	500	< 6	7	0.6-12
Radarsat	SAR	varying	varying	< 6	1	5.3 GHz
Group 2						
NOAA	AVHRR	1100	3000	< 1	5	0.6-12.4
IRS	WiFS	188	770	< 5	2 (3)	0.6-0.9 (-1.8)
Resurs	MSU-SK1	210	700	< 5	5	0.5-12.6
SPOT 4	Vegetation	1160	2250	< 2	4	0.4-1.8
Group 3						
LandSat-5	TM	30/120	185	< 16	7	0.4-12.5
LandSat-7	ETM+	15/30/60	185	< 16	8	0.4-12.5
SPOT 1-3/4	HRV/HRVIR	10/20	60	< 26	3/4	0.5-0.9 (-1.8)

Recently the new American satellite sensor MODIS was launched onboard the Terra-Satellite offering hyper-spectral data in up to 36 bands. This will drastically improve the discrimination of snow and clouds as well as the classification of the snow cover status (dry – wet). In the next few years the new European satellite ENVISAT guarantees the availability of data from different sensors which fly on the same platform. Tab. 2 gives a selection of recently launched and up-coming sensors which can be used for snow cover detection.

TABLE 2. Future satellite sensors for snow cover monitoring.

Satellite	Sensor	Spatial resolution (m)	Swath width (km)	Temporal resolution (days)	Bands	Spectral range (μm)
Terra	ASTER	15/30/90	60	< 16	14/15	0.5-11.7
Terra	MODIS	250-1000	2330	< 2	36	0.4-14.4
Terra	MISR	275-1100	360	< 9	4	0.4-0.9
ENVISAT	MERIS	300	1150	< 3	15	0.4-1.0
ENVISAT	AATSR	1000	512	< 6	7	0.6-12

Figure 1 shows an example of a series of NOAA-AVHRR channels of the Alps and Northern Italy and a digital elevation model of the same area. NOAA-AVHRR data are widely used for snow detection due to its frequent overpasses. They are easily available in most parts of the world at low price level. Due to the spectral range of NOAA-data in the visible, near-infrared, middle infrared and thermal infrared wavelength the discrimination of clouds and snow is mostly possible. Fog in the Po-Valley (A), the Bavarian Pre-Alps (B) and the Swiss Lowland (C) has a different spectral signature compared to the snow-covered Alps or the high clouds along the Croatian coast (D).

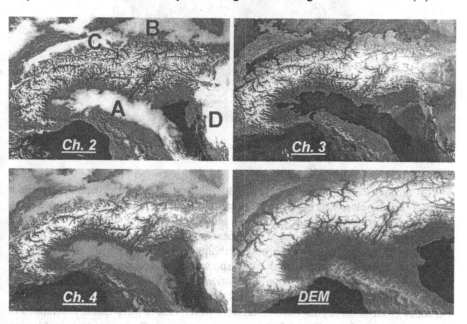

Figure 1. NOAA-AVHRR data of the Alps and Northern Italy: Channel 2 (infrared) : top left, channel 3 (middle infrared) : top right, channel 4 (thermal infrared) : bottom left, digital elevation model (bottom right). A, B, C, D mark the following regions (Po-Valley, Bavarian Pre-Alps, Swiss Lowland, Croatian coast).

Since the vertical depth and duration of snow cover in mountainous terrain is strongly depending on the altitude it is necessary to combine satellite data with a digital terrain model for the analysis. In the meantime digital elevation models (DEM) are available for most parts of the world in a reasonable spatial resolution which fits quite well to the NOAA-AVHRR data. The DEM in Fig. 1 and the DEM of Albania (Fig. 2) are taken from NOAA's GLOBE data set (Global Land One-km Base Elevation Project) which is available on the Internet.

Figure 2. Digital Elevation Model from the NOAA-GLOBE data set showing Albania and Macedonia in a 1-km grid.

For further analysis satellite data and DEM must be co-registered and snow coverage can be classified by using classical digital image analysis tools. Since there will often be a gradient between completely snow-covered and not-snow-covered pixels it is recommended to use fuzzy logic classification techniques to better classify "partly" snow-covered pixels. Through the combination with a DEM statistical analysis are possible to investigate altitudinal shifts of snow coverage or slope and aspect related features (early melting in sun-facing, south-oriented slopes). If very local investigations have to carried out the GLOBE-data set will not be sufficient. In this case a DEM from a national geodetic survey should be integrated which mostly offer spatial resolutions of 10 to 50 m. As an alternative a DEM from the latest Shuttle Radar Topography Mission (SRTM) from late winter 2000 using microwave techniques could be satisfactory. This data set should finally cover Earth's surface between 60° northern and southern latitudes.

3. Methods

3.1. TEMPERATURE INDEX MODELS

Generally one has to assume that satellite data are not available for each day of the snow period (e.g. due to a low repetition period of the satellite system or due to cloudiness during satellite overpass which reduces visibility etc.). Figure 3 shows how satellite can detect snow coverage at certain time slots and how snow cover changes between two satellite overpasses potentially. The temporal change of snow cover is depending on the heat flux into the snow cover and has to be computed by using numerical models. In most cases melting corresponds to an increase of air temperature due to warm air advecion. In this case melting can be parameterised by the spatially distributed air temperature which normally decreases with altitude. One parameter which is used in most of the snow melt models is the degree day factor (Seidel & Martinec 1992, Schaper et al. 1999, Braun 1985, Scherer 1998), which is modified with height. If the heat flux is depending on precipitation this can also be parameterised by a precipitation index factor which describes the sensible heat which is conducted into the snow matrix.

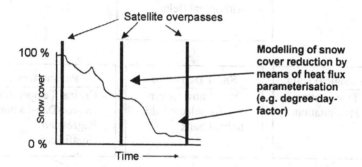

Figure 3. Time slots for the detection of snow cover from satellite data and temporal change of snow cover.

According to Seidel & Martinec (1992) the Snowmelt Runoff Model (SRM) equation reads as:

$$Q_{n+1} = Q_n k_{n+1} + \sum_{i=1}^{N} \left[c_{snow_{n_i}} a_{n_i} (T_n + \Delta T_n) S_{n_i} + c_{rain_{n_i}} P_{n_i} \right] \cdot A_i (1 - k_{n+1}) \frac{10000}{86400} \quad (1)$$

$$\underbrace{}_{\text{Heat related term}} \quad \underbrace{}_{\text{precipitation related term}} \quad \underbrace{}_{\text{catchment size related term}}$$

Q = average daily discharge [$m^3 s^{-1}$]

c = runoff coefficient c_s for snow, c_r for rain

a = degree-day-factor

T = number of degree-days

ΔT = adjustment by temp. lapse rate from the station to average hypsometric elevation

S = ratio of snow covered area to total area

P = precipitation contributing to runoff

A = area of basin or zone

k = recession coefficient indicating the day-to-day decline of discharge in a period without snowfall or rainfall

n = sequence of days

N = number of elevation zones in which the basin is devided

$\underline{10000}$ = conversion from cm• $km^2 d^{-1}$ to $m^3 s^{-1}$
86400

170

All relevant meteorological data are normally available from the official meteorological network and can easily be aggregated according to the requirements of the model. The flow of data which are integrated into the model and the various steps of computation is shown in Figure 4. The model enables simulations of the existing conditions, forecasts e.g. for hydro-power management and simulations related to global climate warming trends.

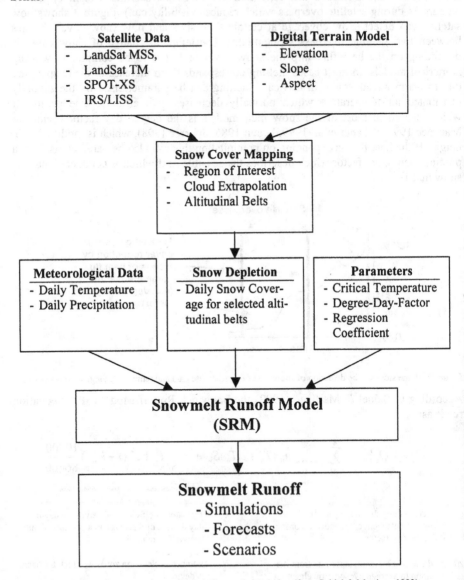

Figure 4. Flow chart for snowmelt runoff modelling and simulation (from Seidel & Martinec 1992).

Under very cloudy conditions net radiation, i.e. the sum of incoming and outgoing short-wave (solar) and long-wave (terrestrial) radiation, can reach values high enough to enable snow melt without very high air temperatures. In this case degree-day-factor models might have a reduced correlation to measurements. An alternative are fully phy-sically based models which simulate to heat balance of the snow cover depending on topography, radiation, advection and snow conditions and the turbulent heat fluxes. This method is very sophisticated and presently is used more on a research rather than an operational level.

3.2. SPATIALLY DISTRIBUTED NET RADIATION

One of the prerequisites for physically based models is the knowledge of the spatially distributed net radiation which is strongly influenced by the surface type and in moun-tainous regions by topography. Combining satellite data, a digital terrain model and ra-diation models it is possible to compute all relevant radiation fluxes and net radiation with great accuracy. There are several methods existing for the final partitioning of net radiation into turbulent heat fluxes. The methodological approach is shown in Figure 5. After some steps of pre-processing (Parlow 1996a) the surface albedo (a) can be calculated from visible and near-infrared channels of satellite imagery. The thermal-infrared channel of LandSat-TM or NOAA-AVHRR is used to compute surface temperatures (T) and the terrestrial, long-wave emission ($E_l\uparrow$). A digital terrain model is the input for a short-wave irradiance model (SWIM) to calculate solar short-wave irradiance on inclined slopes during satellite overpass ($E_s\downarrow$). Finally the long-wave atmospheric counter radiation ($E_l\downarrow$) is estimated by means of a numerical model (COUNTER) on the basis of vertical gradients of temperature and water vapour of the atmosphere. For further information see Parlow (1996b, 1996c).

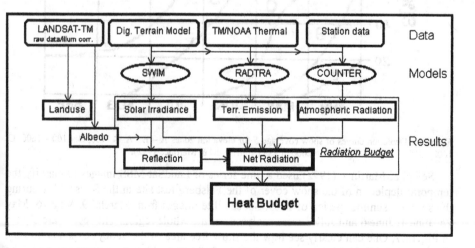

Figure 5. Conceptual model to compute spatially distributed radiation fluxes and net radiation.

Finally net radiation (R_n) can be calculated according to the following equation (with ε : emissivity coefficient and σ: Stefan-Boltzman-constant (5.678 · 10^{-8} W m^{-2} K^{-4}).

$$R_n = E_s \downarrow \cdot (1-a) + E_l \downarrow - \varepsilon \sigma T^4 \qquad (2)$$

4. Results

It is absolutely important to integrate a digital elevation model for the satellite image analysis (Scherer & Parlow 1994). Figure 6 shows an example of the Felsberg region (year 1985) in the south-eastern Swiss Alps which clearly indicates how the decrease of snow cover (S) is delayed depending on the altitudinal belt (A – E) (Seidel & Martinec 1992). At the start of the melting period in early April the lowest region with altitudes less than 1100 m a.s.l. is only 20 % snow covered and snow vanishes until May. In the higher regions B, C and D snow melt begins later in the year and persists until mid-summer and August. In the highest region D (> 2600 m a.s.l.) remains a 20 % snow cover or glaciers during the whole summer. If measurements or good estimations of the snow depth and snow conditions or snow density (dry snow – wet snow) are available for each altitudinal belt it is possible to calculate snow depletion and melt water runoff.

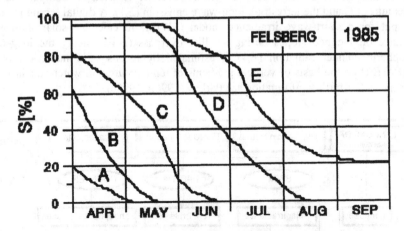

Figure 6. Depletion curves of snow coverage *S* for elevation zones A – E : A: < 1100, B: 1100 - 1600, C: 1600 - 2100, D: 2100 - 2600, E: > 2600 m a.s.l (from Seidel & Martinec 1992).

Seidel & Martinec (1992) used time series of LandSat-MSS images to classify the temporal depletion of the snow cover of the Felsberg test site in the Swiss Alps during the spring – summer period of 1985. Six satellite images from March 29, May 16, May 24, June 1, June 6 and July 3 were analysed for the whole region. The results are shown in Figure 7. One can clearly see how the snow free area of the valleys expands into the slopes and finally only the mountain tops (> 2100 m a.s.l.) remain snow covered. The non-linearity of melting, which is influenced by changes in weather situations (e.g.

warm air advection, colt fronts or rainy periods) is parameterised by the day-to-day temperature course and the precipitation rate according to equation 1. The accuracy of this approach can be seen in Figure 8. The correlation between measured and simulated runoff is very high ($r^2 > 0.9$) and the simulated runoff follows the ups and downs of the weather conditions during the seven months period quite nicely.

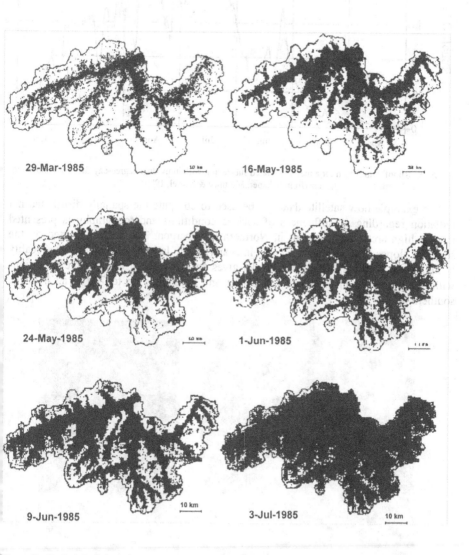

Figure 7. Gradual depletion of snow cover in the Rhine-Felsberg Basin/Swiss Alps for spring 1985 derived from LandSat-MSS data (from Seidel & Martinec 1992). White indicates: snow cover, black indicates: snow-free area.

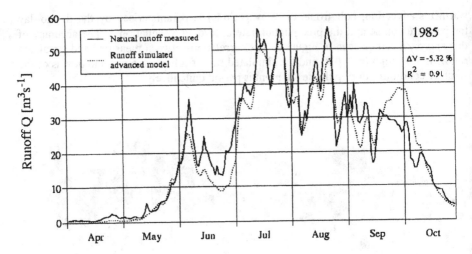

Figure 8. Runoff simulation for a partly glaciated sub-region with individual degree-day-factor for snow and ice (from Schaper, Martinec & Seidel, 1999).

An example how satellite data can be used to compute the spatially distributed net radiation regarding all influences of surface conditions and topography is presented from a high-arctic environment in Northern Spitsbergen. The area is located on the southern slopes of the Liefdefjord area in nearly 80° n. latitude. In Figure 9 the locality in indicated. The digital elevation model gives an overview of topographic conditions with the fjord area in the upper left corner and mountains up to 700 m a.s.l. in the southern part.

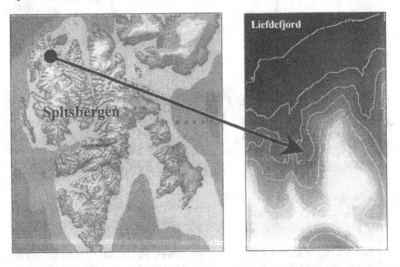

Figure 9. Location of Liefdefjord region in Spitsbergen and digital elevation model with 100 m contour lines.

From the existing digital elevation model (DEM) with a grid size of 20 m slope, aspect and horizon reduction is computed. Then a short-wave irradiance model was used to simulate the solar irradiance during satellite overpass on June 26, 1990 for all DEM pixels with respect to the 3-dimensional orientation of the grid cell.The result is presented in Figure 10. The flat area along the coast line has a solar irradiance of up to 500 W m^{-2}, whereas this is varying from less than 200 W m^{-2} on eastern slopes to more than 800 W m^{-2} on sun-exposed slopes in the southern part of the region. Consequently the topographic conditions have very strong influence on the available solar energy.

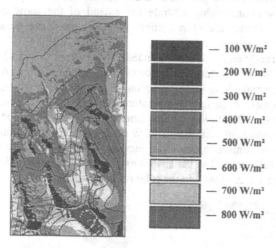

Figure 10. Simulated solar irradiance during LandSat-TM overpass on June 26, 1990 for the Liefdefjord area/Spitsbergen. Please see appendix for image in colour.

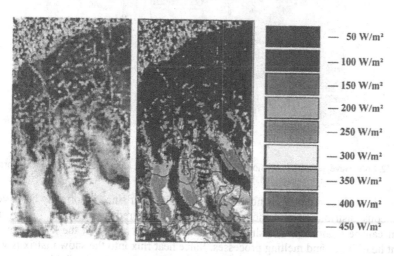

Figure 11. LandSat-TM channel 4 and computed short-wave reflexion during LandSat-TM overpass on June 26, 1990. Please see appendix for image in colour.

After having calculated the short-wave albedo a from LandSat-TM satellite data the short-wave reflexion $(E_s\uparrow)$ can be calculated according to the equation (3). The result is shown in Figure 11.

$$E_s\uparrow = a \cdot E_s\downarrow \qquad (3)$$

Irregular patches of sea-ice are responsible for the relatively high reflection of the fjord surface. Due to scattered snow patches in the lower part of the terrain the reflection is rather small, but with increasing altitude the extend of the snow cover increases, resulting in a general increase of reflection. Highest values can be seen on glacier surfaces and on sun-facing slopes.

Figure 12 is directly analysed from LandSat-TM channel 6 data (thermal infrared) after correction of the atmospheric influence on long-wave radiation. According to the law of Stefan-Boltzmann long-wave emission is related to the 4th power of surface temperature. Therefore, a surface temperature of 0°C (273 K) corresponds to an emission of 316 W m^{-2}. Due to low surface temperatures east-facing and glaciated areas show very low emission of less than 300 W m^{-2}. With increasing altitude there is a decrease in long-wave emission due to an increase of snow cover extend. The small-scale pattern is directly related to the spatial extend of snow-covered/snow-free pixels and – for totally snow-free areas – to the solar radiation input.

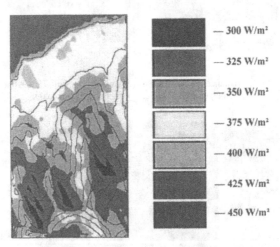

— 300 W/m²
— 325 W/m²
— 350 W/m²
— 375 W/m²
— 400 W/m²
— 425 W/m²
— 450 W/m²

Figure 12. Long-wave, terrestrial emission during satellite overpass on June 26, 1990 for the Liefdefjord area/Spitsbergen. Please see appendix for image in colour.

Atmospheric counter radiation is not displayed in this paper since it normally has a very homogeneous distribution and therefore is often considered as constant. Finally net radiation can be computed according to equation (2). Net radiation the key factor for turbulent heat fluxes and melting processes. Since heat flux into the snow matrix is very small and therefore can be neglected, net radiation balances the available energy for turbulent fluxes and snow melting. As a first guess one can assume that positive amounts of net radiation is nearly completely used for melting at snow covered pixels. Figure 13 shows the spatially distributed net radiation for the time of satellite overpass on June 26, 1990. One can see that the spatial pattern is very complex and a result of multiple reasons. On steep west- and south-facing slopes the input of solar irradiance is the ruling factor. On the other hand there are only minor differences between fjord surface and the nearby shore area. The radiative loss of a high reflectance of the ice-packed fjord and a high emission of the relatively warm shore areas nearly compensate each other in the balance, resulting in nearly similar rates of net radiation. The west-facing slopes in the centre of Figure 13, showing a very high net radiation of up to 450 W m^{-2} are very favourable for a certain type of extreme melting processes, the so-called slush flows (Scherer 1994).

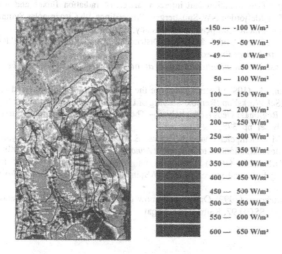

Figure 13. Net radiation of Liefdefjord area during satellite overpass on June 26, 1990. Please see appendix for image in colour.

5. Conclusions

Satellite data play an important role as input for snow-melt runoff models. Data from different platforms are available like LandSat-TM and –MSS, SPOT, IRS-LISS or NOAA-AVHRR. A variety of numerical models is existing, some are already on an operational level while others are still in a research version. Nevertheless, the potential of satellite data for the analysis of snow coverage could be demonstrated. Satellite information could offer an important input variable for snow-melt runoff models. Especially in regions with limited resources for the maintenance of a measurement network satellite data of the various sensor can be used to bridge the gap between cost effectiveness and the need for spatially distributed data.

6. References

1. Braun, L. (1985) *Simulation of snow-melt runoff in lowland and lower alpine regions of Switzerland.* Zürcher Geographische Schriften, 21. Zürich.
2. Gude, M. and Scherer , D. (1998) Snowmelt and slushflows: hydrological and hazard implications. *Annals of Glaciology* 26, p.381-384.
3. Gude, M. and Scherer , D. (1999) Atmospheric triggering and geomorphic significance of fluvial events in high-latitude regions. *Zeitschrift f. Geomorphologie, N.F., Suppl.-Bd.* 115, p. 87-111.
4. Parlow, E. (1996a) : *Correction of terrain controlled illumination effects in satellite data.* In : Parlow, E.: rogress in Environmental Remote Sensing Research and Applications, Balkema Rotterdam, S. 139-145.
5. Parlow, E. (1996b) : *Net Radiation in the REKLIP-Area - A spatial approach using satellite data.* In : Parlow, E.: Progress in Environmental Remote Sensing Research and Applications, Balkema Rotterdam, S. 429-435.
6. Parlow, W. (1996c) : Determination and intercomparison of radiation fluxes and net radiation using LandSat-TM data of Liefdefjorden/NW-Spitsbergen. Proceedings 4th Circumpolar Symposium on Remote Sensing of Polar Environment, European Space Agency, ESA-SP 391, S. 27 - 32.
7. Schaper, J. , Martinec, J. and Seidel, K. (1999) Distributed mapping of snow and glaciers for improved runoff modelling. *Hydrological Processes*
8. Scherer D. (1994) *Slush stream initiation in a high arctic drainage basin in NW-Spitsbergen.* Stratus 1, pp. 96, Basel
9. Scherer, D. and Brun, M. (1997) Determination of the solar albedo of snow-covered regions in complex terrain. - Proc. EARSeLWorkshop 'Remote Sensing of Land Ice and Snow', Freiburg: S. 21-29.
10. Scherer, D. (1998): Regionale Geosystemanalyse – Theorie und Beispiele. Habilitation thesis, Facilty of Science, University Basel, pp. 378.
11. Scherer, D., Gude, M., Gempeler , M. and Parlow, E. (1998) Atmospheric and hydrological boundary conditions for slushflow initiationdue to snowmelt. *Annals of Glaciology* 26, p. 377-380.
12. Scherer, D. and Parlow, E. (1994) Terrain as an important controlling factor for climatological, meteorological and hydrological processes in NW-Spitsbergen. *Annals of Geomorphology, N.F., Suppl.-Bd.* 97, S. 175-193
13. Seidel, K. and Martinec, J. (1992) Operational snow cover mapping by satellites and real time runoff forecasts. Proceedings ISSGH 92, Kathmandu, Nepal

GIS AND SPATIAL DATABASES FOR LAND MINE MAPPING

HELMUT KRAENZLE
Center for Geographic Information Science
James Madison University
Harrisonburg
VA 22807
USA

Abstract

It is estimated that there are over 60 million land mines located in approximately sixty countries no longer involved in wars. These land mines are the cause of up to 25 000 deaths or injuries per year. The majority of victims are civilians working their lands or children playing. Currently there are many governmental, military, and non-profit organisations working to improve the land mine problem. These organisations usually work independently in addressing mine awareness and education, mine removal, and victim relief and assistance in affected countries.

In 1996 the Mine Action Information Centre (MAIC) was founded at James Madison University (JMU) in Harrisonburg, Virginia. The Centre for Geographic Information Science (CGIS) at JMU supports the MAIC with a team of faculty, staff and advanced students.

During the past three years the GIS team has 1) developed customised Geographic Information Systems (GIS) for specific needs of humanitarian demining organisations and operators, 2) evaluated GIS software for a Humanitarian Demining Support System, and 3) hosted an international conference on mapping and GIS for humanitarian demining.

Currently the CGIS is focusing on plans for a clearinghouse for humanitarian demining spatial data. The database clearinghouse will enable the demining community to access information about spatial data sources over the World Wide Web and will provide digital maps.

1. GIS and Humanitarian Demining at James Madison University

1.1. THE CENTER FOR GEOGRAPHIC INFORMATION SCIENCE AT JAMES MADISON UNIVERSITY

The CGIS is within the College of Integrated Science and Technology of JMU. The Centre for Geographic Information Science offers the Bachelor of Science and the

M.F. Buchroithner (ed.),
Remote Sensing for Environmental Data in Albania: A Stragegy for Integrated Management, 179–185.

Bachelor of Arts degrees for a major in geography with concentrations in the following areas:

- Geographic Information Science
- Environmental Studies
- Global Studies

The CGIS has teaching and production labs with 15 Pentium computers, high quality colour printers and plotters, digitising tablets, scanners and a GPS base station. Personal computer software includes PC ARC/INFO, ARCCAD, ARCVIEW, AutoCAD, AutoCAD Map, MapInfo, MapViewer and ERDAS Imagine.

1.2. THE GIS/DEMINING TEAM AT JAMES MADISON UNIVERSITY

The GIS/Demining team is one of six demining teams at James Madison University. The team is a part of the Mine Action Information Centre (MAIC). All team members are either faculty or staff at the Centre for Geographic Information Science (Table 1).

TABLE 1. The GIS/Demining team

Name	Area of Expertise	e-mail address
Dr. Helmut Kraenzle	Team Leader / GIS	Kraenzhx@jmu.edu
Dr. Glen Gustafson	Satellite Image Data	Gustafgc@jmu.edu
Dr. Mary Kimsey	Biogeography	Kimseymb@jmu.edu
Mr. James Wilson	GIS	Wilsonjw@jmu.edu
Dr. Stephen Wright	Digital Mapping	Wrightse@jmu.edu

2. Previous GIS Demining Projects at the CGIS

2.1. REVIEW AVAILABLE COMMERCIAL DATA SOURCES FOR A GEOGRAPHIC INFORMATION SYSTEM FOR CAMBODIA

Under the auspices of the Department of Defence in support of Cambodia, general searches were conducted for country data, aerial image data and map data. The methods of search included extensive use of the World Wide Web, visits to various libraries and centres (e.g., Library of Congress), and extensive telephone interviews. The country data sources were assembled in bibliographic form, that is, as references rather than as hardcopy or softcopy documents. The map and aerial data was first researched with respect to such factors as availability of coverage, date, scale, and cost [1]. The project results were delivered as a final report by the CGIS.

2.2. REVIEW OF COMMERCIAL GIS SOFTWARE FOR A DEMINING SUPPORT SYSTEM

The CGIS obtained detailed information on a wide variety of GIS software packages. This information was categorised and compared to provide the basis for conclusions. A further input to this process was the technical reviews of the various packages, as found in the literature. The basic facts and capabilities of the GIS packages were summarised in a large spreadsheet to provide some overview for judgement [1]. The project results were delivered as a final report by the CGIS. This included a proposed hardware and software architecture for a customised GIS.

2.3. DEVELOPING A CUSTOMIZED GIS FOR A DEMINING SUPPORT SYSTEM

2.3.1. *Using ArcView to Customise a GIS for Humanitarian Demining*
The CGIS developed a GIS prototype including spatial databases for a portion of Bosnia. The GIS was developed using ArcView from Environmental Systems Research Institute (ESRI). ESRI is a dominant GIS software vendor on the world market and ArcView is one of the most widely available GIS software tools in the world.

Using the ESRI programming language, "Avenue", ArcView was used to create a greatly condensed and simplified graphical user interface. When the software wakes up, the customised ArcView screen is visible. It allows certain basic operations to take place almost with the push of one button on the newly designed toolbar (Figure 1, 2). A user can, for example, display any of several different data layers for a desired geographical location. He or she can also take basic measurements from the screen of such things as distances, areas, and ground coordinates.

New geographic features can be added in the field by simply digitising on the screen with the map or aerial imagery in the background. The user can combine the desired layers into a standardised map composition and print it. The amount of training necessary to accomplish all this for a beginner is absolutely minimal, a day or two at most [1].

2.3.2. *Spatial Data used for the Bosnia Customised GIS*
The CGIS acquired Landsat Thematic Mapper data with a resolution of 30 meters for its colour bands, and Spot Panchromatic data with a resolution of 10 meters for its finer spatial resolution. Both satellite images were georeferenced, merged, and integrated into the database. Georeferencing establishes a relationship between image coordinates and known real world coordinates.

Concerning the map data, a 1:50 000 Tactical Line Map from the National Imagery and Mapping Agency (NIMA) was imported as an ADRG file into the GIS. The ADRG raster map is already georeferenced data. The point, line and area features were digitised from the 1:50 000 topographic map and imported as coverages into ArcView. The GIS including the spatial databases were written to a CD-ROM with an installation procedure.

3. The Importance of Spatial Databases for Land Mine Mapping

3.1. SPATIAL DATABASES AND GIS

Spatial databases are an integral part for any GIS and are the basis for mapping and analysing minefields. The most important part of creating a GIS for humanitarian

182

demining is the selection of the appropriate spatial data. In general, a spatial database can be described as data about the spatial location of geographic features recorded as points, lines, areas, or images as well as their attributes (Figure 3).

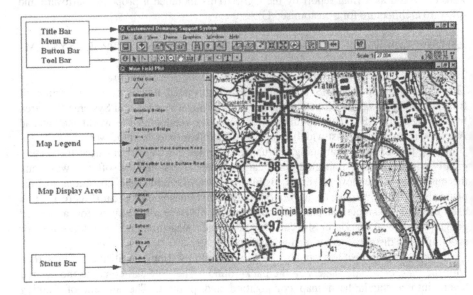

Figure 1. Standard ArcView Interface. Please see appendix for image in colour.

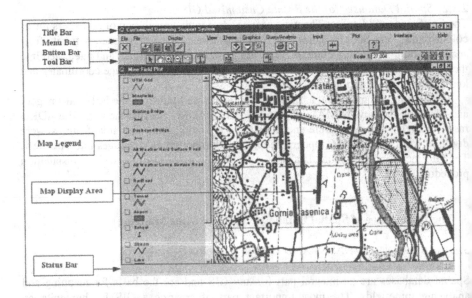

Figure 2. customised ArcView Interface. Please see appendix for image in colour.

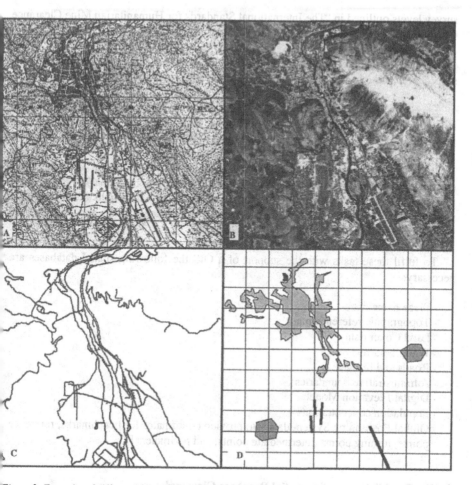

Figure 3. Examples of different layers of a spatial database; taken from the GIS for humanitarian demining in Mostar-Buna, Bosnia
- A) Digital Topographic Map (NIMA 1:50 000)
- B) Satellite Image (Spot Panchromatic & Landsat TM Image Merge)
- C) Road, railroad, and stream data
- D) Built up area, minefields, airports, and UTM grid data

3.2. DATA NEEDS IN HUMANITARIAN DEMINING

The spatial database needed for humanitarian demining is dependent upon the objectives of the demining task. In previous GIS projects for humanitarian demining, the CGIS found, in general, the following digital databases essential: topographic raster maps, satellite images, point features (e.g.; schools, hospitals), line features (e.g.; bridges, roads, streams and tunnels) and area features (e.g.; minefields, airports, built up areas and lakes). For specific demining tasks other spatial databases might be necessary. For instance, certain types of data will be needed for the three different survey levels outlined in "The International Standards for Humanitarian Mine Clearance Operations". These survey levels are defined as follows:

Level 1 – General Survey: "to collect information on the general locations of suspected or mined areas" [2].

Level 2 – Technical Survey: "to determine and delineate the perimeter of mined locations of mined locations initially identified by a level 1 – general survey. The marked perimeter forms the area for future mine clearance operations" [2].

Level 3 – Completion Survey: "to be conducted in conjunction with the mine clearance teams and accurately records the area cleared. The benchmark is to be left in the ground to serve as a minimum marker of the initial minefield area. It is also recommended that permanent markers be used to indicate turning and intermediate points of the perimeter of the mined area" [2].

To fulfil these tasks with the support of a GIS the following spatial databases are necessary:

- Place name data
- Topographic reference maps
- Land Cover data
- Water features
- Roads and bridges
- Administrative boundaries
- Digital Elevation Models
- Population concentrations
- Global Positioning System data – to provide coordinates for benchmarks, reference points, turning points, intermediate points, and perimeters [2].

4. The Establishment of a Spatial Database Clearinghouse

A spatial database clearinghouse is a library of information about spatial data on the World Wide Web. Using metadata, a clearinghouse has the capability to search for the location and availability of spatial data anywhere in the world.

Creating a spatial database clearinghouse specifically for humanitarian demining would be a great benefit to humanitarian demining operations. It would provide the demining community with a central location where spatial databases are available.

Spatial Database Clearinghouse Feasibility Study
James Madison University is currently undergoing a feasibility study to determine the costs and benefits of creating a spatial database clearinghouse for humanitarian demining. The developing components of this clearinghouse includes gathering data, creating metadata and providing general information and support with respect to issues concerning spatial data and GIS. In addition JMU will start to implement a user-

friendly spatial data clearinghouse accessible through the World Wide Web for online users. For users that don't have access to the World Wide Web the information will also be put on a CD-ROM.

The JMU Spatial Data Clearinghouse for humanitarian demining will provide a network for the GIS-Demining community, improving the impact of their operations. The clearinghouse will determine spatial data requirements of Mine Action Centres (MACs) and demining operators and determine the methodology for providing spatial data bases for use in humanitarian demining GISs. Once established, the clearinghouse will support the MACs, NGOs, and others in using GIS for mapping operational areas. The clearinghouse will combine efforts from global demining information centres by acting as a single spatial database, which will be coordinated electronically without being located at one site. It will be a meeting place for the demining community where coordination and sharing of pertinent information will enable different organisations to work together most effectively toward common goals.

JMU will accomplish the following tasks within the year 1999:

- Interview and Survey Key Organisations: Contact relevant agencies and organisations involved with demining to determine their spatial data needs, standards currently in use and major software packages being used.
- Identify Spatial Data Sources: Thorough searches of the World Wide Web, contacting data providers and government agencies.
- Provide Information on Spatial Data Issues and GIS: Provide definitions, brief summaries, examples, and links to published web sites.
- Provide Standards for Spatial Data Sets and GIS: Determine and define current standards used in humanitarian demining for spatial data and GIS.
- Support the Implementation of Standards: Make recommendations of what standards to use and identify major GIS packages used in humanitarian demining.
- Publish Information on the World Wide Web and on CD-ROM: Present results of this study on the World Wide Web and on CD-ROM.

5. References

[1] Gentile, J.; Gustafson, G.; Kimsey, M.; Kraenzle, H.; Wilson, J.; Wright, S. (1997):   Use of Imagery and GIS for Humanitarian Demining Management. In: SPIE Proceedings: The International Society for Optical Engineering, Vol. 3128,   pp. 104-109.
[2] International Standards for Humanitarian Mine Clearance Operations. United Nations. 13 May 1999 <http://www.un.org/Depts/Landmine/Standards/s-   index.htm>.

INFRASTRUCTURE REQUIREMENTS FOR NON-MOTORISED TRANSPORT

Roads and Streets for Animal-based Transportation

EVE IVERSEN
University of California, Davis
1953 22nd Street
San Pablo
California 94806
USA

Abstract

In developing countries animal-based transportation is one form of non-motorised transport that may be critical to the local and national economy. The paper will address how remote sensing systems can be used to evaluate remote areas for improvements that will allow free movement of people and materials. Transportation infrastructures such as roads and bridges are easily distinguished in both air photography and satellite imaging. Pavements such as asphalt retain heat and are also easy to see in thermal imaging. Unpaved roads are not as easily detected but they can be found using high resolution systems. This paper is limited to a discussion of roads and related structures and will not address the infrastructure of other forms of transport.

Animal-based transportation is important in the economies of developing countries. The roads these non-motorised vehicles travel on either are unimproved or have been built exclusively for motor vehicles. This paper will address the needs of animal-drawn transport and how these requirements can be included in road planning. Safety issues such as the causes of runaways and a structure to control this situation will be described. Evaluation of the road user population and a method of conducting a road census will be presented.

1. Introduction

The purpose of transportation is to move people and goods from one place to another. Everything from a person walking to a truck trailer combination moving along a highway fits this definition. Today most engineers and political leaders equate the internal combustion engine with progress. It is assumed that in an age when men have walked on the moon there is no further need for animal-drawn vehicles. For this reason

M.F. Buchroithner (ed.),
Remote Sensing for Environmental Data in Albania: A Stragegy for Integrated Management, 187–210.
© 2000 *Kluwer Academic Publishers. Printed in the Netherlands.*

the infrastructure requirements for this class of transportation are not addressed in planning.

Motorised and non-motorised vehicles both need roads wide enough and strong enough to allow them to move freely. The roads must not have grades or narrow places that will not allow the cargo to be moved. Ideally the road should be passable at all times of the year and in all weather. Bridges and tunnels should be built to allow vehicles to pass without difficulty. The roads should be maintained in good condition all year and be repaired promptly when damaged. It is very rare to find a road that meets all of these requirements anywhere. It is even more rare in countries that are using limited resources to improve the conditions of their citizens. To allow resources to be targeted where they will benefit the most people roads are classified into classes that reflect their importance.

2. Classes of Roads

In developing countries asphalt or other pavement is limited to primary roads. There are several reasons for this. Funds are usually limited and only the primary roads can be completed. Primary roads usually connect major towns and cities and therefore have larger vehicles traveling over them on a regular basis. The primary roads are more easily maintained since their design allows large trucks to get to any part of their length. Finally, primary roads receive more traffic than other roads and need to be kept in the best condition. Primary roads act as channels for imports and exports. They are built primarily to handle motorised traffic and have a few sharp curves and steep grades as the terrain allows.

Secondary roads may or may not be paved depending on the budget of the highway authority. They are usually less well maintained because priority is given to primary roads and secondary roads are given attention only when the primary road work is completed. Secondary roads connect smaller towns with major centres and with each other. They usually carry less traffic than primary roads and the vehicles transiting them are usually smaller. Secondary roads usually act as farm to market roads. Secondary roads usually have sharp turns and steep grades but are passable to most motor vehicles.

Tertiary roads are rarely paved. They may be paths that have been widened or smoothed though this is often not the case. Tertiary roads connect small villages with each other and with towns. High clearance or all wheel drive are usually the only motor vehicles on them. Tertiary roads usually act as postal roads and as a thoroughfare for inter-village trade. Tertiary roads have sharp turns often in the form of switchbacks in mountainous areas and steep grades. These roads are mostly passable to animal-drawn transportation and foot traffic.

2.1. WEATHER FACTORS AND NATO ROAD TYPES

NATO has standardised three descriptions of three types of roads: Type X, Type Y, and Type Z. Each of these types reflects a route's ability to handle maximum traffic capacity in different weather. Weather that can limit the use of a road includes storms, hail, snow, and ice to name only a few. Other conditions such as fog can limit the speed at which vehicles travel but does not effect the traction on the road's surface. The following definition of each route type as given in the US Army Field Manual 5-36 (1970:2-4):

"Type X-All Weather Route is any route which with reasonable maintenance is passable throughout the year to traffic never appreciably less than maximum capacity.
Type Y- All Weather Route (Limited Traffic Due to Weather) is any route which with reasonable maintenance can be kept open in all weather but sometimes only to traffic considerably less than maximum capacity.
Type Z- Fair Weather Route- is any route which quickly becomes impassable in adverse weather and cannot be kept open by maintenance short of major construction."

None of these classifications specifies the type of traffic using the road. It also does not indicate how economically important the road is. Usually a primary road is also a Type X but in severe terrain it may be only a Type Y. This is especially true in low lying areas that are subject to flooding.

3. Remote Sensing and Geographic Information Systems Application to Public Road Information

Remote sensing systems can be very useful in updating the status of a road and rerouting traffic. For example a primary road between two major towns may be obstructed by a landslide. Any vehicle traversing the road would either have to turn back at the obstruction or risk forcing away through. A vehicle that turns back delays the delivery of passengers or cargo. If the vehicle forces a way through there is a risk that it will either become stuck or be involved in an accident. Either way there is a loss in time and often money that could have been prevented by the delivery of timely road conditions information.

Using remote sensing a central or regional highway department can monitor road conditions in remote areas. In conjunction with weather forecasters it is possible to predict which areas of the road will become blocked and to have road crews standing by. For example, if a primary road crosses a stream that floods rapidly after a storm in its watershed it will be possible to predict that the crossing will be unusable until the waters subside. A road crew can be dispatched to the crossing to put up signs and warn traffic to take another road. The crew can also insure that the waterway does not become dammed with debris and thereby make a bad situation worse.

Remote sensing combined with a geographic information system can also be used to monitor earthquake damage to roads, bridges, and tunnels. After a major shock remote sensing instruments can view roads in areas known for slope failure and determine if work crews need to be dispatched. The geographic information system can help in correlating the list of damaged sites and help the road department determine repair priorities. Remote sensing also allows road authorities to survey large areas without risking their staff members. Finally, remote sensing provides a "big picture" of damage and allows engineers and political leaders to determine how much time, money, and effort will be needed for a specific area to recover.

4. Non-motorised Transportation

Non-motorised transportation includes pedestrians, hand carts, bicycles, tricycles, equestrians, and animal-drawn transport. For this paper I will use horse-drawn vehicles as my example. The general information that will be presented is applicable to all animal-drawn transport.

Roads and streets laid out before 1930 in most places were designed with the needs of pedestrians, motorists, and animal-drawn transport in mind. Primary roads built after World War II (1939-45) were usually built for motorised traffic alone. Non-motorised transport attempting to use these primary roads often has to get out of the lane or off the right of way entirely. The situation becomes even more dangerous on bridges and in tunnels where there is little or no room to escape. In developing countries many modern roads were designed by engineers who had no understanding of the types of non-motorised traffic that would be using the right of way. Without knowing what types of vehicles needed to travel on the road the construction was geared to support trucks and cars only.

Secondary roads were improved if the development budget allowed. These roads may not have been paved and were more likely to be widened, straightened or reduced in gradient by the highway department. Tertiary roads were usually not improved unless there was a political or military reason to do so.

Today most engineers have no training, background or experience in building multi-user roads. No textbooks give tables for width required by animal-drawn vehicles or the radius of a turning circle for a large wagon. None give a description of the design of safe road shoulders that animal-drawn vehicles can use to escape traffic. There is no material still in print that describes modifications to urban bridges to prevent accidents caused by runaway teams. For rural areas there is no information on how to evaluate the needs of small villages for new or rebuilt tertiary roads when the primary traffic is non-motorised.

5. Basic Principles of Animal-drawn Transportation

The effectiveness of animal-drawn transportation is more difficult to calculate than mechanised transportation because the strength of any one animal or team can vary on a day to day basis. Variation can also occur due to the state of repair of the vehicle and the road surface. The figures given in these tables are based on engineering experiments and can only approximate the situation that will be found in the field.

Animal-drawn transportation is different from mechanised transportation in one fundamental factor: animals get tired. Table 1 shows the relationship between the speed an animal is asked to work and the amount of load he can pull. Level iron rails and a level macadam road are the standards of comparison. The resistance in both cases is considered the same but the speed of the horse is increased. If an animal is asked to move faster both the load and the duration of work must drop. If a load has to be moved quickly, having relays of horses at intervals to take over is necessary as each team tires. This was the system used for postal couriers and stage coaches.

TABLE 1. Relationship between Speed, Hours of Work and Load for Horse-Drawn Transportation. (Byrne 1907: 443 converted to metric)

Speed in Kilometers per Hour	Duration of the Day's Work hours/minutes	Resistance to Traction kilograms (assumed)	Useful Effect of One Horse working 1 day in tons drawn at 1.61 km.*	
			On Level Iron Rails. Tons.	On Level Macadam. Tons.
3.62	11:30	37.88	115	14
4.83	8:00	37.88	92	12
5.63	5:54	37.88	82	10
6.44	4:30	37.88	72	9
8.05	2:54	37.88	57	7.2
9.66	2:00	37.88	48	6.0
11.27	1:30	37.88	41	5.1
12.88	1:23	37.88	36	4.5
14.48	0:54	37.88	32	4.0
16.10	0:45	37.88	28.8	3.6

* The actual labor which a horse can perform is greater, but he is injured by it.

TABLE 2. Tractive Power of Horses at Different Velocities . (Byrne 1917: 7 converted to metric)

Kilometers per Hour	Tractive Force (kg force)	Kilometers per Hour	Tractive Force (kg force)
1.20	151.20	3.62	50.40
1.61	113.40	4.02	45.36
2.01	90.72	4.42	41.24
2.41	75.60	4.83	37.80
2.81	64.80	5.63	32.40
3.21	56.70	6.44	28.35

TABLE 3. Duration of a Horse's Daily Labor and Maximum Sustained Velocity .
(Byrne 1907: 442 converted to metric)

Duration of Labor, Hours	Maximum Velocity, Kilometers per Hour	Duration of Labor, Hours.	Maximum Velocity, Kilometers per Hour
1	23.66	6	9.66
2	16.74	7	8.85
3	13.68	8	8.37
4	11.75	9	7.89
5	10.62	10	7.40

Tables 2, 3, and 4 addresses the trades off between speed, force and hours of work. Table 2 demonstrates as the speed increases the tractive force of the horse decreases. Table 3 shows that velocity drops as the hours of labor increase. Table 4 confirms that more tractive force can be presented when the number of hours of work is decreased.

TABLE 4. Increase in Tractive Power with Decrease in Work Time.
(Byrne 1907:442 converted to metric)

Hours Per Day	Tractive Force (kg force)	Hours Per Day	Tractive Force (kg force)
10	45.36	7	66.61
9	50.35	6	75.45
8	56.70	5	90.72

As a rule of thumb from 1.20 - 6.44 kph the power is inversely related to speed. Another rule of thumb is that to get the greatest amount of work for the longest time from a horse without hurting it you must not exceed one eighth the maximum amount the animal can do for a brief time. These points become important when loads have to be moved over less than ideal roads.

Table 5 shows that two horses in a team cannot pull twice the load one horse can. Combined they can pull 190% of what one horse can pull under the same conditions. This is important when considering using larger teams to pull large loads. There is never a 100 percent increase in tractive value as each horse is added. Therefore improving the road is better than to keep adding horses.

TABLE 5. Tractive Power of Teams in Terms of one Horse (Byrne, 1917:8) .
The tractive power of teams may be found by multiplying.

Number of horses	tractive value per horse	tractive value of team
1	1.00	1.00
2	.95	1.90
3	.85	2.55
4	.80	3.20

Tables 6, 7 and 8 are related to road grades. A horse's power is used in part to overcome the effect of gravity due to his own weight and the weight of the load. In addition there is the factor of how much traction can the animal get on the surface. The smoother a level surface is the easier it is for the horse to move a load. Once the ascent begins a smooth road is not advantageous because the horse cannot get enough foothold to push against as it steps up and forward. The road must be smooth enough for the wheels of the wagon to turn easily but the horse must be able to thrust against it and lean its shoulder into the collar. The force is then transferred from the collar to the load and the load is moved. If the horse cannot get enough traction against the pavement, he cannot pull the load even if he has the strength to do so.

TABLE 6. Loads in kg that a Horse Can Draw upon Various Surfaces and Grades.

Kinds of Surface.	Rate of Grade.							
	Level.	1 %	2%	3%	4%	5%	10%	15%
Earth road-Good	1361.00	1088.62	907.20	725.75	635.03	544.31	362.87	136.08
Poor	589.67	498.95	408.23	317.51	272.16	226.80	181.44	68.04
Broken-stone-good	1814.37	1224.70	907.20	725.75	635.03	544.31	317.51	90.72
poor	725.75	498.95	362.87	272.16	226.80	204.12	113.40	45.36
Stone Blocks-good	2721.55	2041.17	1496.90	1224.70	997.90	771.11	408.23	181.44
poor	1361.00	1043.26	771.11	635.03	498.95	408.23	204.12	90.72
Asphalt-clean & dry	4536.00	1814.37	1133.98	816.47	589.67	453.59	181.44	----

TABLE 7. Data for Loaded Vehicles over Inclined Roads.
(Byrne 1917:11 converted to metric)

Rate of Grade %	Equivalent Length of Level Road (km)	Maximum Load Horse Can Haul (kg.)
0 %	1.61	2844.02
0.25 %	1.80	2438.51
0.50 %	2.00	2255.71
0.75 %	2.21	2036.63
1.00 %	2.41	1880.14
1.25 %	2.61	1737.26
1.50 %	2.81	1625.68
1.75 %	3.01	1492.32
2.00 %	3.22	1412.49
2.25 %	3.41	1331.29
2.50 %	3.61	1236.04
2.75%	3.80	1188.41
3.00 %	4.00	1127.63
4.00 %	4.80	994.83
5.00 %	5.54	816.47
6.00 %	6.41	711.23
7.00 %	7.80	620.06
8.00 %	8.02	560.19
9.00 %	7.79	510.29
10.00 %	9.62	467.20

TABLE 8. Comparative Rank of Pavements (Byrne 1917:184).

Characteristics	Variety									
Qualities	Value (per cent)	Asphalt(sheet)	Asphalt (block)	Concrete	Macadam (bituminous)	Macadam (water-bound)	Brick	Granite	Sandstone	Wood
Low tractive resistance	20	20.0	19.0	18.0	19.0	11.0	18.0	12.0	14.0	20.0
Service on grades	10	3.0	3.0	7.0	4.0	8.0	9.0	10.0	10.0	2.0
Non-slipperiness	5	1.5	2.5	4.0	2.5	4.5	3.5	3.5	5.0	2.0
Favorableness to travel	5	5.0	4.5	3.5	4.0	4.5	3.5	3.5	4.0	4.5
Sanitariness	10	10.0	9.0	7.0	8.0	3.0	8.0	6.0	7.0	9.0
Noiselessness	3	2.5	2.5	2.0	2.5	2.5	1.5	1.0	1.5	3.0
Minimum Dust	3	2.5	2.5	2.0	2.0	1.0	2.0	1.5	2.0	2.0
Ease of Cleaning	5	5.0	5.0	3.5	4.0	1.0	3.5	1.5	1.5	5.0
Acceptability	4	3.5	3.5	2.5	3.0	1.5	2.5	2.0	2.5	4.0
Durability	15	7.5	8.5	6.0	3.0	1.5	10.0	15.0	14.0	11.5
Ease of maintenance	5	3.5	4.0	3.0	3.0	2.5	4.0	4.5	5.0	5.0
Cheapness (first cost)	10	4.5	4.0	5.0	7.5	10.0	4.0	3.0	3.5	3.0
Low annual cost	5	1.5	2.5	3.0	3.5	1.0	4.5	5.0	5.0	5.0
Totals	100	70.0	70.5	66.5	66.0	52.0	74.0	68.5	75.0	76.0

As Table 8 shows if the load is equal and only the road surface is compared the smoothest road needs the fewest horses to pull a ton of cargo. Asphalt is the best pavement for level roads. If it has to be used in mountainous areas the grade should be minimised by the construction of switchbacks. Stairs can also be built into the pavement. The tread should be inclined forward and down and be at least 20 cm deep. The risers should be no more than 20 cm tall. The risers must be as equal in height and the treads as equal in width as possible. Stairs are useful for pedestrians as well. Motorists and bicycles have a problem with stairs and if the road will be used by mixed groups of users the stairs should be set to the side or in the centre and built so that they will not damage wheeled vehicles.

All weather dirt roads are the best roads in the mountains. In urban areas other factors such as dustiness, sanitariness, and easy of cleaning should also be considered. In rural areas the all weather nature of the surface and its traction are the most important factors. The road should be as free from dust in dry weather and mud in wet weather as possible.

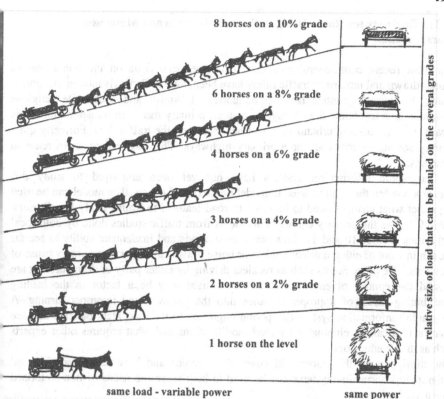

8 horses on a 10% grade

6 horses on a 8% grade

4 horses on a 6% grade

3 horses on a 4% grade

2 horses on a 2% grade

1 horse on the level

same load - variable power

same power

relative size of load that can be hauled on the several grades

Figure 1. The load one horse can pull on the level is equal to the load eight horses can pull on a 10% grade. (Redrawn from Chatburn 1921: 26 by Leah Patton)

General Rules for Animal-based Transportation

1. A small load can be moved quickly for a long period of time
2. A large load can be moved quickly for a short period of time.
3. Over a long distance either decrease the speed or decrease the load.
4. Avoid inclined roads if possible even if it requires a longer distance.
5. On inclined roads take the smallest grade even if it requires a longer distance. Increased grade means that more work is expended in fighting gravity than moving the load.
6. When paving inclined roads provide the best traction possible.
7. When paving level roads traction is important but other factors can be considered.
8. Infrastructure such as tunnels and bridges must be designed to take the special needs of animal-based transportation into consideration. These needs include proper turn-out areas to avoid traffic, wide lanes that allow motorised and non-motorised traffic to avoid each other and methods of stopping runaways.

7. Traffic Safety for Roads Using Both Animal-drawn and Motorised Transportation

The most recent comprehensive studies that have been done on the interaction of animal-drawn and motorised traffic safety have been done in Pennsylvania, in the United States. In several counties large communities of Amish and Mennonite religious denominations use horse drawn buggies as their primary means of transportation. This area is also undergoing urbanization that has increased the traffic flow. Formerly quiet country secondary roads are now primary highways. This has lead to an increase in accidents.

Geographic information systems have not yet been employed to study this phenomenon but they can be used to model this series of events. The model can be used to predict what changes need to be made to road features to insure safety for all users. Table 9 shows the primary causes of accidents from traffic studies done by both local authorities. Tables 10 and 11 show that narrow roads and inadequate ability to see far enough in front of either a motor vehicle or horse drawn carriage is the primary cause of accidents. The other causes such as reckless driving by either party or drunk driving are beyond the control of engineering. Lines of sight may be a factor in the leading contributing factor of improper entrance into the roadway and improper turning. A geographic information system can permit engineers to track accident patterns and see what factors can be eliminated by road modifications and what requires other experts such as driver educators.

The third part of this paper will cover these topics and how remote sensing and geographic information systems can be used to help serve the needs of non-motorised transportation

TABLE 9. Effect on Grades upon the Loads a Horse can Draw on Different Pavements and Grades, Byrne (1907:445).

Grade	Earth	Broken Stone	Stone Blocks	Asphalt
Level	1.00	1.00	1.00	1.00
1 %	.80	.66	.72	.41
2 %	.66	.50	.55	.25
3 %	.55	.40	.44	.18
4 %	.47	.33	.36	.13
5 %	.41	.29	.30	.10
10 %	.26	.16	.14	.04
15 %	.10	.05	.07	.....
20 %	.04	.....	.03	.....

TABLE 10. Number of Horses Required to move one Ton on Different Pavements.
(Byrne 1907:6)

Asphalt 1.00
Stone blocks, dry and in good order .... 1.50 to 2.00
Stone blocks, in fair order .... 2.00 to 2.50
Stone blocks, covered with mud .... 2.00 to 2.70
Macadam, dry and in good order .... 2.50 to 3.00
Macadam, in a wet state 3.30
Macadam, in fair order 4.50
Macadam, covered with mud 5.50
Macadam, with the stones loose .... 5.00 to 8.20

TABLE 11. Contributing Causes of Accidents (Blame not Assigned), (Non-Motorised Vehicle Study Lancaster County Planning Commission February 1993:22).

Contributing factor	Number of accidents
Improper entrance	28
Careless passing	20
Following too closely	15
Drunk driving	11
Failure to obey traffic device	9
Improper turning	8
Speeding	6
All other causes	17

TABLE 12. Mid-Block Horse and Carriage Accidents. Non-Motorised Vehicle Study Lancaster County Planning Commission February 1993:22.

Accidents on narrow roads	43
Accidents on adequate roads	25
Total number of accidents	68

TABLE 13 Road Intersection Horse and Carriage Accidents. Non-Motorised Vehicle Study Lancaster County Planning Commission February 1993:23.

Intersection with poor sight lines	34
Intersection with adequate sight lines	12
Total intersection accidents	46

Runaways

nimal-drawn transportation shares the road with motor vehicles in most parts of the
orld. As the number and speed of cars, trucks, and busses increases the number of

accidents with slow-moving carts and wagons also goes up. Often during road construction it is possible to plan for the needs of both classes of traffic and widen the shoulders so that the slower moving vehicle can yield the right of way. If an animal or team panics there is a place to slow them down and regain control.

Bridges are narrower than roads and rarely have places where teamsters can pull out of traffic to calm frightened horses. If the driver cannot literally "hold his horses" (get the animals under control) the horses will panic, and a runaway will ensue. A frightened team can collide with, or run over, other vehicles or pedestrians. When horse-drawn vehicles and motor vehicles share bridges, the problem is even more direct.

Since there is very little current experience with the problem of runaway horses in urban areas I have gone back to the engineering reports from the early 20th Century for a solution. This problem was recognised in the United States at the beginning of the 20th Century and a unique engineering solution was developed. The simple system of a pair of wooden gates in the shape of a "V" and an understanding of equine psychology helped save lives in the past. The same system can save human and animal lives on bridges today.

8.1. CAUSES OF RUNAWAYS

Horses are prey animals that escape predators by running away rather than fighting. They will spook at the slightest unfamiliar sound, smell, or sight. Edison Monthly [Magazine] (1912) reported the results of the runaway accident investigations on New York City bridges for the twelve months spanning April 1909-April 1910 as follows; " Of the 42 runaways which occurred . . . 14 were caused by broken harness and 22 from unknown causes. Under this heading are probably classed all of those in which the animals were frightened by the noise of passing trains and trolley cars or automobile horns, steamboat whistles and bad driving. Starting from anywhere out on the span [Williamsburg Bridge], the animals go down a three percent grade and attain terrific speed by which time they reach the [barricade] . . . Into which the horses crashed at full speed. [The barricade] was effective in stopping runaways, ending lives and destroying properties." (Edison Monthly).

8.2. THE RUNAWAY PROBLEM IN NEW YORK CITY

During the late 19th and early 20th Centuries, runaway teams of horses dashed across the bridges spanning the East River of New York City almost daily. The Brooklyn Bridge was built in 1883 and spans the East River from Manhattan to the then independent city of Brooklyn[1]. The Williamsburg Bridge connects the East Side of Manhattan with the Williamsburg section of Brooklyn by spanning the East River. It was opened in 1903.

In the early 1900's, an accident occurred which caused the public to bring pressure of the city authorities to find a way to stop bolting teams. The incident was caused when a

---

[1]Brooklyn was an independent city until 1898 and was then incorporated as a borough of New York City.

pair of horses panicked at the sound of a ferry boat whistle, while crossing the Brooklyn Bridge. The team jerked the driver from his seat and then ran across the bridge. The team and wagon sped into Manhattan's crowded Park Row and killed a woman and her baby. In response to the public outcry a "Runaway Barricade" was approved for installation on major bridges. The Brooklyn and Williamsburg Bridge were each fitted with a Runaway Barricade stop stampeding horses.

8.3. THE RUNAWAY BARRICADE

The Runaway Barricade was a 3 meters (ten feet) tall moveable wall made of timber. It was normally kept tied to the side of the bridge. The Runaway Barricade was intended to stop panicked horses at all costs. The bridge was under constant police surveillance. At the sound of a runaway alarm, a policeman swung a simple wooded wall from its storage position parallel to the side of the bridge's traffic way, into a position perpendicular to traffic. Any horse and rider, or any vehicles from a bicycle to a horse-drawn wagon would slam into the wall if the driver could not stop in time. This included anyone whom the barricade pinned in front of the oncoming runaway.

The Runaway Barricades were effective in stopping panicked horses but most of the time the horse, or horse team was killed. Frequently the driver was seriously injured and the wagon was destroyed. The cost in lives and property was too high and again the municipal authorities of Manhattan and Brooklyn looked for a better way to stop panicked horses from racing over the bridges. The bridge authorities needed a better solution.

8.4. DESIGN, INSTALLATION AND OPERATION OF THE RUNAWAY GATE

James Connors, a worker on the Williamsburg Bridge came up with a device was to be composed of two leaves (halves), each just under 2.13 meters (seven feet) tall and 12.2 meters (40 feet) long. The base of the "V" was as wide as the roadway. The leaves would be moved from their storage position parallel to the sides of the bridge, into an angled position that formed a "V" with an open apex about 30 cm (one foot) wide. The base of the "V" faced the direction of traffic. The magazine Edison Monthly (1912) gave a detailed description of the mechanism:

"There are six push buttons at regular intervals along the bridge span, as soon as the horse panics the nearest policeman runs to his booth and sends the signal. At the sound of an electric bell, the bridge patrolman [at the gate] leaps to his place at the controller, ready to close the gates. Here the officer must exercise his coolest judgment, for right in the path of the runaway may be another wagon that must not be trapped. A shout adds speed to the progress of this rig, and just as it dashes past, the current is turned to the motors and slowly the big gates come closed.

A 7-horsepower electric motor, operating a worn gear and rack and pinion, to, which s, attached a steel arm, pushes each of the leaves into position. The mechanism was installed by the Bridge Maintenance Company and is inspected three times a day.

Figure 2. Front view of the Runaway Gate on the Williamsburg bridge in New York City.
(Reprinted from Popular Electricity 1910: 414)

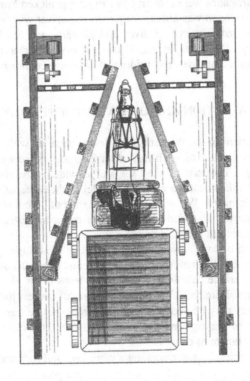

Figure 3. Overhead view of the Runaway Gate on the Williamsburg bridge in New York City.

When a horse finds himself blocked on all sides, he makes for the opening. He is
then suddenly stopped. He does not scrape his shoulders, or even come in contact with

the sides of the chute. The wagon does all of that. On the steel sheathing, there are long scratches. That is where the hubs of more than 100 runaways have struck. As the chute narrows, the grip on the hubs tightens and in a distance of from 1 foot to 8 feet (.3-2.4 meters), according to the angle at which the team enters the chute, the whole outfit is at standstill.

It takes six seconds to bring the leaves of the gate into position, and as the leaves cease their motion, the runaway horse at full speed dashes into the chute and comes to an abrupt stop. The narrowing sides of the chute gripped the wagon at the front hubs and brought it to a stop long before the horse reached the points where he would have come in contact with the chute walls. All the checking power had been applied from the rear and the animal found himself shouldered with a burden he could not move. It was a simple operation, and the horse escaped with barely a scratch. After a few seconds, in which he [the horse] works off his surplus energy and calms down to his ordinary self, the hold is released and he is driven away."

Once the horse had calmed down the policeman assigned to the bridge released the leaves of the "Runaway Gate" were returned to their original position parallel to traffic. When the Gate was opened, the pressure came off the vehicle's axle and the horse could continue to pull the wagon across the bridge.

Popular Electricity reported: "In one runaway which occurred since its installation, the horse started 150 feet (45.7 meters) distant, and by the time it had reached the gate, the leaves were nearly closed. The horse breaking away from the harness passed through the opening without injury, leaving the wagon at the apex of the V. The leaves swing back parallel to the sides of the bridge when not in service."

8.5. VALUE OF THE RUNAWAY GATE

The "Runaway Gate" was put into operation on April 14, 1910. It was successful in stopping every panicked horse that entered it. There were no injuries to either animals or people. None of my sources gave the comparative construction costs for the "Runaway Barricade" and the "Runaway Gate" but two sources did provide casualty figures.

TABLE 14. Comparison of the Runaway Barricade vs. The Runaway Gate. (based on Edison Monthly 1912 and Popular Electricity 1910)

Runaway Barricade						
Dates	Runaway Numbers	Horses Involved	Runaway Stopped	Human Dead	Human Injured	Horses Dead	Horses Injured
1906-10	185	246	127	not given	93	53	47

Runaway Gate						
1910-12	100+	not given	all	none	none	none	none

9. Environmental Considerations

9.1. WINTER ROAD ICING

It is now customary to apply salt to city streets and primary roads to prevent ice from accumulating or to remove it once it has formed. Salt works by lowering the freezing point or water and usually acts to turn ice into water. The exact temperature where freezing will occur is depends on the concentration of salt. The wet streets can then absorb heat during the day and the warmed water will then evaporate from the surface. At night if the temperature drops below the freezing point for that salted area a saline slush will be formed which is colder than an unsalted section. The saline slush has bad effects on shoes and boots made of leather as well as the feet of animals especially if they are made to stand in it for hours.

The feet of all equines and bovines becomes chilled in the extra cold slush. The sensitive lamina is chilled and the animal will pick up its feet alternately to try to relive the pain. Any open wound that contacts salt causes intense pain. If the animal stands in the saline slush for protracted time the heel will become ulcerated. The hoof will rot and crumble causing the animal to become totally unfit to work.

There is an alternative to salting where animal-based transportation is used: apply sand to the road. Sand will not cause the problems that salt does to the feet of people and animals. Another reason for not applying salt is that salt runoff will pollute streams and kill roadside vegetation. Sand will not pollute and it can be shoveled out of areas where it is a nuisance.

9.2. DRY WEATHER DUST

Dirt roads are often dusty in dry weather. In some areas this is controlled by applying oil. This is not a good system because oil runoff can pollute creeks and poison soil. It is better to cover roads with gravel which is not as prone to blow when it is dry. Gravel can also improve drainage from the surface and is easier to shift into ruts and holes. The disadvantage of gravel is that it provides more resistance to the wheels of vehicles thereby slowing traffic. Smaller vehicles following large trucks can also get gravel kicked into their windshield when trucks drive too fast on this surface.

9.3. WET WEATHER MUD

Mud is the wet weather face of dust. It is the worst surface for any person, animal, or vehicle to have to travel on. It not only reduces traction but it saps the strength of people and animals as they fight for a foothold. Constant use of an unimproved road will often produce a mud bath when rain comes. Depending on the consistency of the morass the amount of work it takes to lift one foot can be two or more times what it is in any other type of substrate. This reduces the amount of distance that can be covered and the load that can be moved.

Both animals and humans can become totally mired if the mud is above their "knees". They have to be pulled out if a firm enough footing can be found to work from.

Legs can slip out from under causing a fall. Bruises, strains, sprains, and broken bones can result. If the hind legs of an animal go out laterally in a "split" dislocations or a broken pelvis can occur. Fighting the clinging mass of wet soil quickly exhausts all species. Weaker individuals may suffer heart attacks. Collapse risks drowning in the quagmire.

The only way to prevent a road from becoming a bog is to build a good foundation with proper drainage. The more time and money that are spent on the road bed the longer the road will last and the less it will cost to maintain it. One example of the longevity of a well built road system are the imperial Roman roads. These highways have lasted over two millennia and can still be used. Goods roads are emblem of the confidence a nation has in its future.

10. Traffic Census

A traffic census provides data for determining which roads need to be placed on a priority list for construction or reconstruction. There are many ways to make a census including the use of automated systems but most of these are not designed for use with anima-drawn transportation. The system outlined below is based on one developed by Captain F.V. Greene and conducted early in the 1900's. It was designed only for horse-drawn traffic and has been adapted for use with other species using the abstract from Bryne (1907:19-20)

Greene Animal-drawn traffic census

Hours and days of observation: For six work days (to be determined by local working customs) in the same location, from 0700-1900 except when darkness prevents observation.

TRAFFIC CENSUS   Page One

of ....Street

.................................City..............State.......

Date.............................

Observer:..

Weather..

....

Temperature at two hour

intervals:..

Class of pavement:

dirt..........asphalt........concrete.........brick.........cobblestones......... NATO

classification..............

Condition:

dry........damp..........icy.........snow.........dusty........greasy.......rutted.......potholed.....go

od.......

Width of road (curb to curb)..................... Drainage

(describe)..

Shoulder (sketch road below, measure width, and describe construction)

TRAFFIC CENSUS   Page Two

Classification of Vehicles and loads	Hours of Observa- tion						
	0600- 0700	0700- 0800	0800- 0900	0900- 1000	1000- 1100	1100- 1200	1200- 1300
1 donkey cart light							
1 donkey cart heavy							
1 horse/mule cart light							
1 horse/mule cart heavy							
1 ox/buffalo cart light							
1 ox/buffalo cart heavy							
2 donkey wagon light							
2 donkey wagon heavy							
2 horse/mule wagon light							
2 horse/mule wagon heavy							
2 ox/buffalo wagon light							
2 ox/ buffalo wagon heavy							
Large donkey teams (enter animal count)							
Large horse/mule teams (enter animal count)							
Large ox/buffalo teams (enter animal count)							
Led animals							
Number of falls *							

Remarks
* Note under remarks if the fall was on knees, haunches, or complete and the cause if possible

TRAFFIC CENSUS   Page Three

Classification of Vehicles and loads	Hours of Observation					
	1300-1400	1400-1500	1500-1600	1600-1700	1700-1800	1800-1900
1 donkey cart light						
1 donkey cart heavy						
1 horse/mule cart light						
1 horse/mule cart heavy						
1 ox/buffalo cart light						
1 ox/buffalo cart heavy						
2 donkey wagon light						
2 donkey wagon heavy						
2 horse/mule wagon light						
2 horse/mule wagon heavy						
2 ox/buffalo wagon light						
2 ox/ buffalo wagon heavy						
Large donkey teams (enter animal count)						
Large horse/mule teams (enter animal count)						
Large ox/buffalo teams (enter animal count)						
Led animals						
Number of falls *						

Remarks
* Note under remarks if the fall was on knees, haunches, or complete and the cause if possible

TRAFFIC CENSUS   Page Four

Guidelines for estimating weights of vehicles:
Less than one half ton
1 donkey single axle cart

Less than one ton
1 horse or mule single axle cart empty or lightly loaded
1 horse or mule two axle wagon empty or lightly loaded
1 horse or mule one or two axle carriage
2 donkey two axle wagon loaded

Between one and three tons
1 ox or water buffalo cart lightly loaded (closer to 1.5 tons)
2 oxen, water buffaloes wagon fully loaded (2 tons)
1 horse or mule cart heavily loaded (closer to 1.5 tons)
1 horse or mule wagon heavily loaded (closer to 2 tons)
2 horse or mule wagon heavily loaded (closer to 3 tons)

Over three tons
Wagons pulled by teams of more than two oxen, water buffaloes, horses, or mules and
fully loaded. This can also apply to wagons carrying very heavy cargo such as broken
stone or metal bars.

Figure 4. Barrels of beer traveling over good roads in Europe. (Reprinted from the magazine Good Roads)

Figure 5. Bulky hay traveling over good roads in Europe. (Reprinted from the magazine Good Roads)

Figure 6. Quagmire on a street in the early 1900's in the United States. (Reprinted from the magazine Good Roads)

Figure 7. Rutted rural road in the early 1900's in the United States. (Reprinted from the magazine Good Roads)

11. References

1. Byrne, Austin T. (1917) *Modern Road Construction*. American Technical Society. Chicago, Illinois.
2. Byrne, Austin T. (1907) *A Treatise on Highway Construction*. John Wiley and Sons. New York.
3. Chatburn, George R.(1921) *Highway Engineering; Rural Roads and Pavements*. John Wiley and Sons. New York.
4. Editor, *Edison Monthly Magazine* (1912) Stopping the Runaway. August. 1912. 78-80.
5. Lancaster County Planning Commission (1993) Non-Motorised Vehicle Study .February 1993. Lancaster County, Pennsylvania. USA.
6. Maul, Norman (1912) New Way to Stop Frightened Horses *Technical World*. November. 1912. 348-349.
7. Editor, *Popular Electricity* (1910) "Bridge Gate to Stop Runaways". September. 1910. page 414.
8. US Army (1970) *Rout Reconnaissance and Classification* Field Manual 5-36

POTENTIALS AND LIMITATIONS OF TECHNOLOGY TRANSFER IN THE DEVELOPMENT CO-OPERATION, SHOWN FOR THE TRANSFER OF INFORMATION TECHNOLOGY

Appropriate Information Technology Transfer: A Contribution to Development

GERHARD BECHTHOLD, BERTHOLD HANSMANN
Gesellschaft fuer Technische Zusammenarbeit (GTZ)
Division 4500
P.O. Box 5180
D-65726 Eschborn
Germany

"Access to Information is Access to Development"

Abstract

The Paper shows that Information Technology (IT) is the driving force of a large part of current developments both in industrialised countries and in the developing world. Transfer of know-how and experiences is essential.

Transfer through technical co-operation has to be done with consideration of the target environment, i.e. the conditions in which the technology shall be applied: The technology has to be transferred in an *appropriate context* in order to fit to the available resources and needs, and thus to be *sustainable*.

With a successful IT transfer, developing countries can participate at the benefits of IT, can create jobs, increase the prosperity and *through use and access to knowledge and information* can *improve the efficiency and transparency of their political systems*.

1. Background

1.1. CURRENT GLOBAL CHANGES AND DEVELOPMENT

Technology transfer takes places every day and everywhere in the development co-operation. It is one of the main objectives of development assistance to have *developing countries mastering technologies*, and to apply them to improve the living conditions and political freedom of their people.

This article shall elaborate on some aspects, potentials as well as constraints of transfer of the current Information Technology (IT) phases to the Developing World.

The World Development Report 1998/99, published by the World Bank, says that 'in an increasingly knowledge-based economy, information is becoming at least as

211

M.F. Buchroithner (ed.),
Remote Sensing for Environmental Data in Albania: A Strategy for Integrated Management, 211–219.
© 2000 *Kluwer Academic Publishers. Printed in the Netherlands.*

important as land and physical capital. In the future, the distinctions between developed and non-developed countries will be joined by distinctions between fast countries and slow countries, networked nations and isolated ones.'

This statement reflects the current change of technology appliances and economies. There are different opinions about the scope, values and consequences of such present changes, there are different views and assessments about the advantages and disadvantages, but no one can argue that such a change is currently taking place. It is becoming 'inevitable'.

Many aspects of our *daily life are changing,* not only in the

- *Technology* as the basis - the way, how we communicate, move, work, live, travel, but also in
- All aspects of our society - the way, we spend our time, we make and value personal and professional experiences, we live together, we form political units etc.

Information Technology is becoming a backbone of our new society; information dissemination is shaping the societies in many countries. Currently, we are living in a transition from a 'traditional' production society to an *'information society'*, or 'IT era' or 'post-industrial society', which is characterised by *domination of services* over other economic sectors and *niche instead of mass markets.* By definition, these characteristics can serve in form of *potential benefits for* small, not necessarily centralised economies, e.g. of *small developing countries.*

Results of this new formation can be seen everywhere in the industrialised countries:

Abundant global information flow, media, supply of information, global networking of service functions. This also implies the use and access to knowledge and information *with transparencies, political awareness, freedom of expression, as well as processing of data, availability of services, know-how and education. Presently, these are on such an advanced level, as never have been imagined before.*

In line with the strong value of information and the handling of *information as a catalyst to development,* there is an accompanying change in the fields of economical and political liberalisation and deregulation. Companies are being re-structured, 're-engineered', there are global fusions of large companies, international trade is booming: *Globalisation* combines the appearance of technology advances with global liberalisation.

- Information is capital!
  α   Human resources are becoming more important - as 'resources'.
- Distance is becoming strongly irrelevant for many aspects!
  α   Site-specific is loosing its traditional value; network technologies do not make a difference between a few and thousands of kilometres.
- Traditional definitions of development discrepancies with definitions of industrialised and developing countries will slowly disappear.
  α   There will be a 'South' in the 'North', as well as a 'North' in the 'South'.

This implies a number of considerations for future planning and co-operation from the technical point of development assistance in order to share the resources and prosperity equally among the people of the Earth.

Public *in industrialised countries* is well aware of these phenomena. Discussions about 'computers' and the Internet show an increasing interest and participation in the ongoing changes. People become conscious about the future 'digital era'. Public and private institutions are taking actions to cope with these changes.

1.2. DEVELOPING COUNTRIES: SITUATION AND AWARENES

Developing countries are heterogeneous in their characterisation. Take-off countries such as some Far Eastern countries have already made big progress, while others, mostly LLDC (least developed countries), still have substantial infrastructure and political problems.

They can not be left behind! To exclude them and not to transfer technology, would mean not to let them benefit. The gap between the situation in the 'North' and the 'South' would widen - with further 'un-development' in those countries in Africa, Latin America, Asia, for which already now a lesser development is characteristic.

In most developing countries a strong awareness exists, that currently in the industrialised countries a change and technology rush occurs. This consciousness might even be stronger than in European countries.

About 80% of the world's population is still unable to keep pace with the revolutions in the computer and communications world. But, some 40% of the world can catch up in one way or another. It is the fate of the remaining 40% of the world that is hanging in the balance.

Such a view is not a single view of a newspaper publisher, but an often-heard opinion in many developing countries, no matter which technology level they are in. Developing countries have to play part in the movement of technology application.

Over the past couple of years this has increasingly been led to the formulation of organisations and institutions, which try to define and improve the impact of current global changes to the Developing World. Just to mention a few:

- The Global Knowledge Partners
- Commonwealth Network of Information Technology for Development
- Africa ONE
- Pan Asia Network, bringing the Internet to rural villages (e.g. in Bangladesh)
- Grameen Bank, which uses Internet facilities to break the cycle of poverty, and which uses 'micro loans' to fund 'village pay phones'

including activities by World Bank, United Nations and many other institutions.

One of the most obvious and most discussed new features with the strongest impact is the *Internet*. While business in industrialised countries can not be imagined anymore without Internet and email connections, also developing countries show a strong increase in the use of Internet, despite some critical comments of conventionalists. New technologies, like wireless networks ('local loops'), digital networks, fibre-optic cabling, and financing schemes enable more people, including the poor, to have access to the Internet.

But the Internet is only one component of our new IT era. It shall therefore not be overemphasised in this review.

2. Objectives and Approaches of IT Development Assistance

The focal point is not, if the trends and globalisation are 'good' or 'bad', but to recognise the trends and to *stress on activities, how to make best use* of the benefits of the current, global changes and developments.

To transfer technology and know-how through technical co-operation in the field of IT, revised strategies both in 'North' and 'South' have to be defined:

- *Industrialised countries* have to open their trading and know-how policies. Liberalisation can be a promising activity into a fruitful direction. Many NGOs and bilateral aid agencies are active to improve the situation. 'Developed countries should understand the necessity and democratic right of the poorer countries to gain access to the information superhighway' (N.Mandela).
- *Developing countries* show a different level of preparedness for getting into the next millennium with different governmental, society and technology structures. Some countries do not have the awareness and see development only as a grass-root level activity, where only the basic need of people have to be fulfilled. These countries require a strong input for awareness building.
- Other countries are very dedicated and prepared to step into new mechanisms to participate in technology and to be able to compete with industrialised countries *in specific aspects.* A well-known example is the technology development in take-off countries in the Far East.

The IT transfer has therefore to be defined according to the level of the developed country.

In the long-term, IT transfer to developing countries has to be seen in two ways:

- The *passive role* as *a recipient* is their position to receive transferred know-how, training, with systems to be installed and an awareness building, with a flow of IT from 'North' to 'South' through technical co-operation by bilateral institutions, NGOs, or private sector.
- For an *active role* at a later stage they have to be prepared and given the opportunity for *active integration* in the global 'IT sector', which will be a decentralisation of IT affairs, procedures and work tasks.

It is well possible to integrate IT development components *within the current frame of technical assistance*!

Technical co-operation in projects is normally well defined with goals, objectives and terms of references. Activities and tasks are oriented on the level of implementation and project execution.

With considerations above, an *additional impact of projects* can be in the frame of IT development assistance. It can consist of the transfer of know-how to developing countries for an *'institution building for IT* (and IS)' in order to enhance the awareness and capabilities in information system-related fields and activities: A core group is to be trained, information systems to be set up, information policies and data access policies to be defined and implemented.

These *can be 'side-effects'* or spin-off achievements of technical co-operation projects, or it can be accessed through a defined strategy in co-operation with local authorities and projects.

3. IT Projects and Activities

Commonly, the general term IT is being used. Indeed, it would be better to differentiate between IT, for 'information technology', and IS, for *'information system'*. The latter *includes aspects of management, institution building*, awareness building, integration in government structures, user approval etc. These are crucial components, when a system is being set up, or training is given.

Projects working on development and co-operation in the field of IS or with an IS component are experiencing different phases, as seen from the long-term IS perspective:

IS is being transferred to a country in different phases:

- At the beginning, when IT is still rather virgin in an application field in a country, any activities and discussions are rather seen as 'exotic'. *No awareness* has been created yet.
- The next phase is often characterised by *'computer experts'* in institutions or projects with *all responsibilities given to*, without involvement of the management.
- This is often followed by an *over-emphasis of the IS advantages* and facilities, in *technology-driven projects.*
- In a long-term view, the final step of this IS implementation cycle is a *realistic consideration* and understanding (plus integration) of the IS, where information is seen as a *tool*, often playing only a 'secondary role'.

This is an often-observed long-term *cycle of awareness development of the value and reputation of information systems*. There is hardly any possibility to avoid this learning curve. Only the last phase can be seen as a success, while the previous stages are still 'failures', but necessary phases of a development cycle.

On the other side, it is well possible - even within a project - to *stress the importance of management integration in the technological aspects* and to target the IS set-up activities towards the defined long-term goals.

In practical terms, transfer of IT in the technical co-operation can consist in following fields, with partly overlapping activities:

- IS set-up, application oriented, including marketing and management aspects,
- Management training and awareness building among 'decision makers' and managers,

- Transfer of (technical) know-how, training in the fields of ICT (information and communication technologies), hardware, software, programming,
- Small-scale or 'informal' IT dissemination efforts , such as set-up of small training and educational organisations, which can quickly adapt to change, small training centres for the youth ('call them cyber cafes or youth clubs or whatever'),
- Infrastructure improvement (technical, funding): telecommunication lines, electricity
- Formal and informal training on trainers,
- Propagation of (easy) access to information, Internet access,
- Loan, small-scale provision of funds for procurement of IT facilities,
- Establishment of low-cost Internet providers,
- Demonstrations, shows, fairs, workshops.

4. Appropriateness and Sustainability

Project experiences show that *sustainability* of transfer activities is *highly correlated with the appropriateness* of introduced and transferred know-how. Therefore, highest priority has to be given at the set-up of a project to the *resources assessment* and evaluation of the environment, in particular to the *human resources assessment*. If done correctly, the project will have a good chance for success, i.e. for:

- Sustainability,
- Dissemination and
- Replication for other similar tasks in the country.

All actions are to be considered less under the aspect of performance, but *under the aspect of appropriateness*. Considerations can vary for many aspects:

- Installed *hardware* can range from stand-alone PCs with standard configuration to complex mainframe systems,
- *Operating systems*: from standard, GUI based Windows versions to UNIX systems,
- *Software*: from simple general- purpose programs to specific applications for highly skilled staff,
- *Application level*: from simple word-processing to complex applications in modelling, communication, statistical algorithms,
- *Interfaces*: from step-by-step wizard-guided menu-driven systems to command-line level, need for interface programming and conversion between different programs and data formats,
- *Communication*: from without external communication to completely networked systems, Internet, intranets,
- *Integration in institutions* and (government) agencies: from stand-alone application in a very specific set-up to full integration in policy and development institutions,
- *Training*: from few staff members to thousands; institution- internal or external; from training courses of a few days to sophisticated programming and hardware maintenance courses,

- *Compatibility*: from stand-alone to full compatibility between many institutions, full support and/or development of data exchange formats.

Beside hardware limitations such as computers, network, communication lines, electricity, availability of computer and spare parts, it is particular the question, to which degree the application should be developed *for easy use*, and where and by whom this application shall be prepared.

Western academic and performance standards are less relevant than sustainability. Donors are becoming more aware of this fact, while vendor companies still try to catch customers with the latest, often expensive, but rather inappropriate versions of hardware and software. This will often result in failures.

In the field of technical co-operation, potentials of IT can be exploited - and success will be given a high chance -, if the transfer is *under appropriateness clearance*. It might be well possible to think of an (informal or formal) *'IT Appropriateness Certificate'* for the acquisition or approval of related projects.

Potential in the transferred IT field and success of the project is high, if the approach is multifunctional, with integration of technical *and management aspects* and if it is *adjusted to the local environmental conditions* and to the local know-how of the staff and their understanding and cultural approach. These aspects have to be integrated in an assessment. A 'human resources assessment' is often the key to define the appropriateness in relation to the staff and therefore the software and automation level of the IS to be installed.

Limitations are given - and failures will occur -, if IT is being simply copied form industrialised countries without considering the specific needs, environment, local human resources and infrastructure.

5. Perspectives

This article reflects about project implementation aspects, which are important for the definition and execution of technical co-operation projects, as done e.g. by GTZ. In the following, a perspective shall be given to the future possibilities and *benefits of successfully transferred IT*.

There can be many benefits for the developing countries, if IT is transferred successfully and human resources are developed. Following are some of the benefits:

- *Outsourcing of IT work*: Both rather advanced tasks (e.g. programming, application development) as well as labour intense jobs (e.g. data entry, accounting, call-centres, monitoring security screens, performing on-line services) can be *outsourced* from companies in industrialised countries to developing countries. They can then form a resource of *income* of individuals, *create jobs* and contribute to *a better balance of trade*.
- *Creation of off-shore IT sectors* (e.g. services, banking/financing)
- *Global diffusion of knowledge*: Access to information, which was once restricted to the industrial world and travelled only slowly beyond it.
- *Stop of emigration* of skilled workers, as poor countries with good IT will be able to retain their human resources

- Utilisation of the Internet to serve the needs of grassroots organisations
- Impact on good governance and politically stable, democratic developments, based on informed citizens
- Potential to *'leapfrog' to new technological* stages with enhanced infrastructure, making the use of IT more economical and more efficient
- Creation of new markets
- Narrowing the development and knowledge gap between industrialised and developing countries
- Education
- Telemedicine
- *Opportunity to global peace*, as countries become more economically interdependent, global trade and foreign investment will grow and people will communicate more freely.

Today, this is already practised in a few countries such as India, East Europe, Russia, Caribbean region etc. In future, with further globalisation trends and with an ever-stronger Internet, this will become more important. It might give a chance to countries to survive, *particularly countries where natural resources are limited, but with a 'demographic potential'* of people with a relatively high education and awareness level and good international languages skills. They can benefit from those 'niche instead of mass markets'.

It shall be noted that IT is an *extremely dynamic field. Developments in (hardware) technologies* as well as in the progressive field of *appliances* (i.e. user friendliness and integration of IT in the every-day life) are very fast. Pace of technological development is likely to accelerate, not to slow down.

- There will be no end of the need of governments, individuals and enterprises *to adjust to new IT techniques*, both in industrialised and in developing countries. It will depend very much on the successful IT transfers during the next 10 years, how far transfer to developing countries will still be still required in the long term- or if the countries can learn themselves by having achieved a high level of awareness and trained cadre of personnel.
- Technical and detailed prognoses, analyses and forecasts can only be valid for a short period. *More important are long-term policies*, optimistic and constructive attitude, visionary concepts and the ability to *restructure and to adjust to the new situation.*

The term 'IT' is the general umbrella, under which many individual activities exist. One of the specific application fields, which deals with spatial distributions, is *GIS*: Geographical Information Systems. All what have been said about IT and the transfer of IT is obviously valid for GIS transfer in the same way.

Future of the use and importance of GIS is very bright. The technology of GIS has reached a high level of usefulness. It has been shown, that a well-defined and set up system can contribute to development and co-operation in the context of *appropriate IT transfer.*

GIS and all Information Systems will revolutionise all aspects of living and planning, including all development aspects. But this will be for the better only if their output is geared closely to practical objectives and is continuously validated.

6. Outlook

'Globalisation of markets and economies, fuelled by electronic networking, presents tremendous challenges to developing countries. But, globalisation and the information revolution present no threats, but hopes and opportunities. They give the developing world a dramatic chance to leapfrog into the future, breaking out of decades of stagnation and decline.' [after A. Fatoyinbo, D+C 2/1999]

In the technical co-operation, let us share this viewpoint, accept the challenge and work actively for the *'hope and opportunities'*.

MANIFESTO OF TIRANA - RESOLUTION OF NATO REMOTE SENSING & GIS WORKSHOP 1999

1. The NATO Workshop on Remote Sensing and Geographic Information Systems held in Tirana from 6 to 10 October 1999 took notice of the economic and technological transition problems Albania is presently facing.

2. It recognised, that there exist strong initiatives to overcome present deficiencies in the country by a strategic approach using modern technologies to combat the existing problems of sustainable development of the country. This regards both infrastructure, agricultural and environmental degradation to strengthen the national economy on a multidisciplinary basis.

3. If established, the new technological advances in satellite remote sensing, digital mapping and geographic information systems can serve as a basis to improve the country's economic and environmental development conditions, provided they are adapted to local requirements.

4. The Workshop participants recommend that a common multidisciplinary approach for the utilisation of these new technologies should be accomplished by the following steps:

4.1. Creation of a geo-information working group as a platform for information exchange with the aim of later establishing a sort of agency for the co-ordination of multipurpose geo-information activities.

4.2. The aim of this working group could further extend:
   a) to develop a national action plan for utilisation of remote sensing and GIS and to facilitate the establishment of centres for these multidisciplinary activities.
   b) to request hands-on training workshops for Albanians in remote sensing and GIS through donor funding.
   c) to initiate meta-information exchange of existing geo-data which are currently held by isolated institutions and not commonly utilised.

5. The Workshop participants acknowledge the tireless efforts of Prof. Ergjin Samimi and his Albanian colleagues to realise this meeting as a first initiative and thanks NATO for funding the meeting.

Tirana, 10 October 1999

M. Buchroithner.

Prof. Dr. Manfred Buchroithner
Workshop Director

LIST OF SPEAKERS

WOLFGANG BAETZ
Gesellschaft für Angewandte Fernerkundung mbH (GAF)
Arnulfstrasse 197
80634 Munich
Germany
e-mail: *baetz@gaf.de*

TOMAS BENEŠ
ÚHÚL Forest Management Institute,
Nábrezní 1326
250 44 Brandýs nad Labem,
Czech Republic
e-mail: *benes@uhul.cz*

MANFRED F. BUCHROITHNER (Workshop Director)
Institute for Cartography
Dresden University of Technology
Mommsenstr. 13
D-01062 Dresden
Germany
e-mail: *buc@karst9.geo.tu-dresden.de*

GEORGE BÜTTNER
FÖMI Remote Sensing Centre
Bosnyák tér 5
Budapest
H-1149 Hungary
e-mail: *buttner@rsc.fomi.hu*

PAOLO CIAVOLA
Dipartimento di Scienze Geologiche e Paleontologiche
Università degli Studi di Ferrara,
Corso Ercole I d'Este 32
44100 – Ferrara
Italy
e-mail: *cvp@dns.unife.it*

224

NINA D. COSTA
Strategy and Systems for Space Applications Unit
Space Applications Institute
European Commission
TP 261
Joint Research Centre
I-21020 Ispra (VA)
Italy
e-mail: *nina.costa@jrc.it*

BERTHOLD HANSMANN
Gesellschaft fuer Technische Zusammenarbeit (GTZ)
Division 4500
P.O. Box 5180
D-65726 Eschborn
Germany
e-mail: *berthold.hansmann@gtz.de*

JOACHIM HILL
Remote Sensing Department
University of Trier
D-54286 Trier, Germany
e-mail: *hillj@uni-trier.de*

EVE IVERSEN
University of California, Davis
1953 22nd Street
San Pablo
California 94806
USA
e-mail: *ehiversen@ucdavis.edu*

GOTTFRIED KONECNY
Institut fuer Photogrammentrie und Ingenieurvermessung
University of Hannover
Nienburger Str. 1
30167 Hannover, Germany
e-mail: *gko@ipi.uni-hannover.de*

HELMUT KRAENZLE
Center for Geographic Information Science
James Madison University
Harrisonburg
VA 22807
USA
e-mail: *kraenzhx@jmu.edu*

ILIA KRISTO
Faculty of Economic
University of Tirana
Rr. e Elbasanit
Tirana
Albania
e-mail: *ikristo@usa.net*

ALFRED MOISIU
Albanian Atlantic Association
Bul. "Deshmoret e Kombit"
Pallati I Kongreseve
Tirana
Albania
e-mail: *amoisiu@abissnet.com.al*

EBERHARD PARLOW
Institute of Meteorology, Climatology and Remote Sensing
University Basel
Spalenring 145
CH-4055 Basel
Switzerland
e-mail: *eberhard.parlow@unibas.ch*

MYSLYM PASHA
Military Topographic Institute
Rr. "M. Keta"
Tirana
Albania

226

NIKOLAS PRECHTEL
Institute for Cartography
Dresden University of Technology
Mommsenstr. 13
D-01062 Dresden
Germany
e-mail: *prechtel@karst8.geo.tu-dresden.de*

PERIKLI QIRIAZI
Department of Geography
University of Tirana
Tirana
Albania
e-mail: *genti@softhome.net*

ERGJIN SAMIMI (Workshop Co-Director)
Alb-Euro Consulting
Rr. "Jani Vreto" Nr. 29
Tirana
Albania
e-mail: *esamimi@icc.al.eu.org*

THIERRY TOUTIN
Canada Centre for Remote Sensing
588 Booth Street
Ottawa
Ontario
Canada, K1A 0Y7
e-mail: *Thierry.Toutin@CCRS.NRCan.gc.ca*

APPENDIX

AVAILABILITY OF CURRENT SPACEBORNE EARTH OBSERVATION DATA

Figure 1. Natural colour composite of IRS-1C-Pan/LISS data with 5m resolution, showing the airport of Tirana (Albania). Copyright: ANTRIX/SII/euromap 1999

230

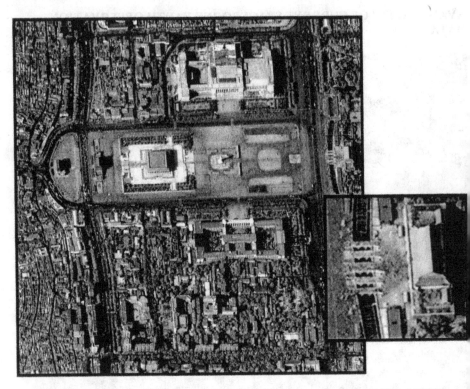

Figure 2. 1m resolution IKONOS data showing a part of Beijing (China). Copyright: SI 1999, GAF 1999

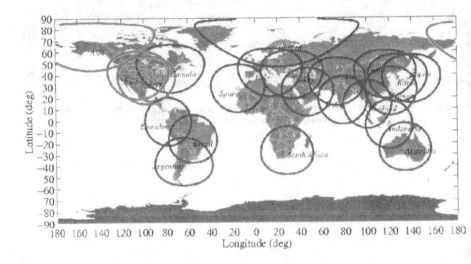

Figure 4. Space Imaging network of international ground stations.

WWW INFORMATION SERVICES FOR EARTH OBSERVATION AND ENVIRONMENTAL INFORMATION

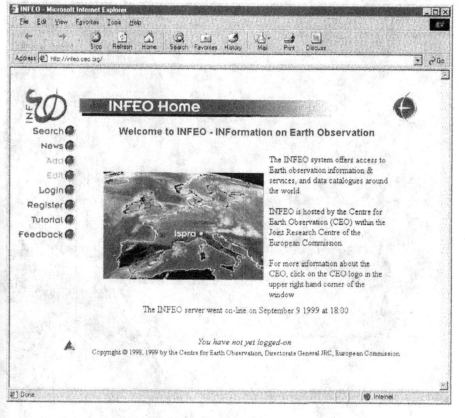

Figure 2. Screen shot of the INFEO home page.

MAP MAKING WITH REMOTE SENSING DATA

GEOMETRIC CORRECTION

Figure 2. Comparison of two composite subortho-images (4 by 3 km; pixel of 5 m) by the photogrammetric method (top) and by the polynomial method (bottom), to which the road vector file (accuracy of 3-5 m) has been registered. The radiometric processing performed are the same for both images.

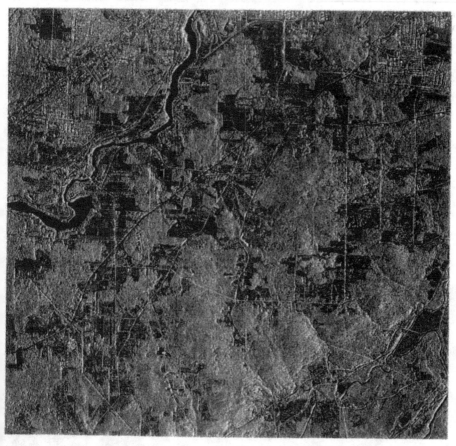

Figure 4. Composite ortho-images (10 by 10 km; 5-m pixel spacing) using IHS radiometric coding with SAR-west in I, SPOT-P in H and SAR-east in S.

COMPUTER-ASSISTED LARGE AREA LAND USE CLASSIFICATIONS WITH OPTICAL REMOTE SENSING

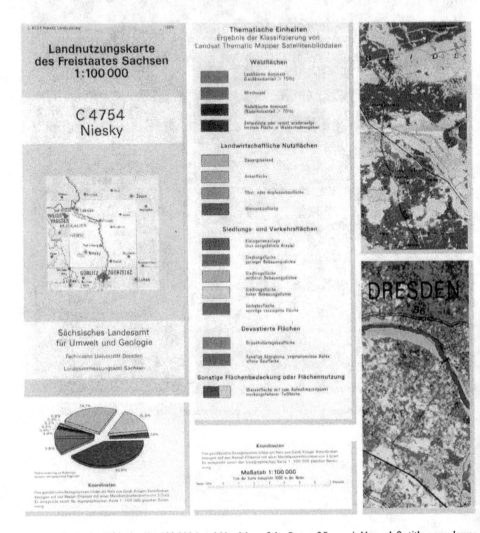

Figure 7. Design of Series '1 : 100 000 Land Use Map of the State of Saxony'. Upper left: title page, lower left: pie chart of class frequencies, upper middle: thematic legend, lower middle: coordinates and scale, upper right: section of map face from sheet 'Niesky' in reduced scale, lower right: section of sheet 'Dresden' in original scale.

RESOURCE ASSESSMENTS AND LAND DEGRADATION MONITORING
WITH EARTH OBSERVATION SATELLITES

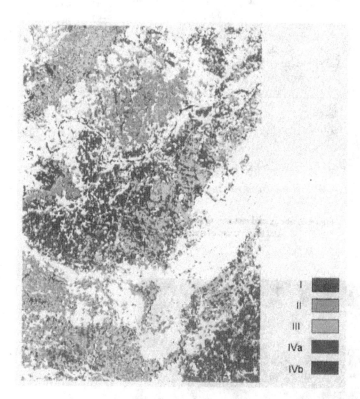

Figure 5. Soil condition map for the Ardèche study site, obtained from the spectral mixture analysis of Landsat-TM data: white areas have an estimated green vegetation cover above 50 % and were not analysed in terms of their soil properties.

Figure 7. Estimates of green vegetation abundance (i.e., proportional cover), resulting from the use of fixed (a) and spatially adaptive (b) endmember sets in a study site with Mediterranean species (Ardèche, France) (from Hill et al, 1995).

236

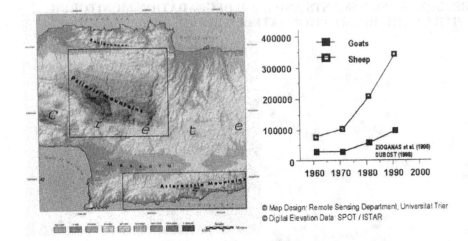

Figure 8. Rangeland monitoring sites in the mountainous ecosystems of Crete, and the increase of ruminants in the Psiloriti region.

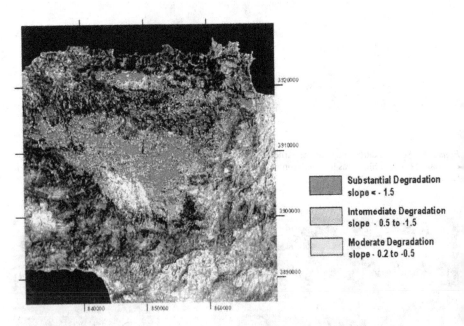

Figure 11. Trend analysis of vegetation cover for the period 1977-1997, based on Landsat-MSS and -TM time series covering the Psiloriti highlands above 1000 m. The colours indicate areas where degradation processes led to a significant loss of perennial plant cover; areas with stable (or partially increasing) vegetation cover are not colour-coded.

COASTAL ZONE GEOMORPHOLOGICAL MAPPING USING LANDSAT TM IMAGERY: AN APPLICATION IN CENTRAL ALBANIA

False colour composite image obtained combining bands 1 and 5 of the 1986 and 1996 scenes. The three insets are images of the Shkumbini (a), Semani (b) and Vjosë (c) deltas obtained by subtraction of equivalent bands between the more recent and the older scene. The areas marked in green indicate land loss, the areas in red land gain.

SNOW RUNOFF MODELS USING REMOTELY SENSED DATA

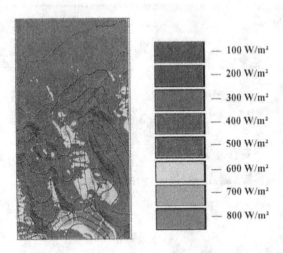

Figure 1. Simulated solar irradiance during LandSat-TM overpass on June 26, 1990 for the Liefdefjord area/Spitsbergen.

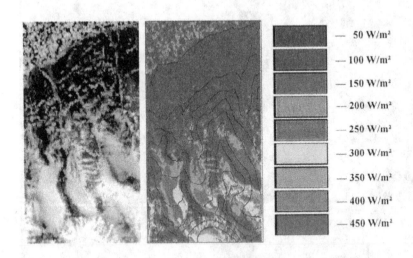

Figure 2. LandSat-TM channel 4 and computed short-wave reflexion during LandSat-TM overpass on June 26, 1990.

239

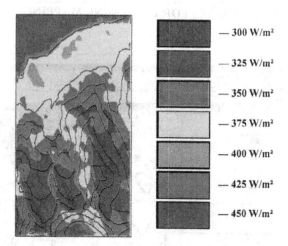

Figure 3. Long-wave, terrestrial emission during satellite overpass on June 26, 1990 for the Liefdefjord area/Spitsbergen

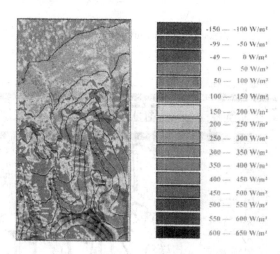

Figure 4. Net radiation of Liefdefjord area during satellite overpass on June 26, 1990

GIS AND SPATIAL DATABASES FOR LAND MINE MAPPING

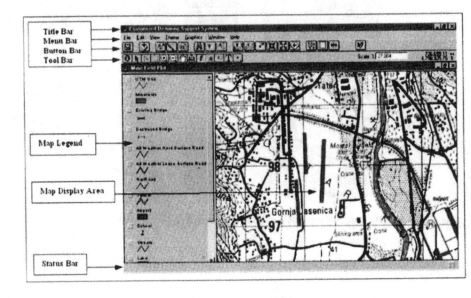

Figure 1. Standard ArcView Interface

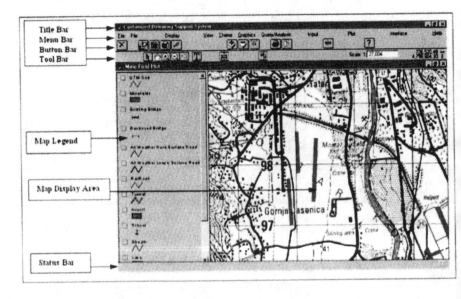

Figure 2. customised ArcView Interface.

SUBJECT INDEX